D1832940

WEATHER

Owing to the antiquity and rarity of this book, certain pages are regrettably absent, and readability is compromised due to the blurring of the original text. If you possess any insights regarding the availability of these missing sections, kindly inform us. Your assistance is crucial in ensuring the completion of this significant work for the benefit of future generations.

WEATHER

A POPULAR EXPOSITION
OF THE NATURE OF WEATHER CHANGES
FROM DAY TO DAY

BY THE

HON. RALPH ABERCROMBY

GYAN

Published by

GYAN

Gyan Publishing House
5, Ansari Road
Daryaganj, New Delhi-110002
Phone: 011-47034999, 9811692060
E-mail: books@gyanbooks.com

Distribution Network
gyanbooks.com
India, USA, Canada, UK, Australia

© Publisher

All rights reserved. No part of this work may be reproduced, stored, adapted, or transmitted in any form or by any means, electronic, mechanical, photocopying, micro-filming, recording or otherwise, or translated in any language, without the prior written permission of the copyright owner and the publisher.

ISBN: 978-81-212-8458-5 (HB)
First Published, 1888

2nd Impression 2024

Printed at: Gyan Press, Delhi.

The book is sold subject to the condition that it shall not, by way of trade or otherwise, be lent, resold, hired out, or otherwise circulated without the prior publisher's written consent.

WEATHER
Author: RALPH ABERCROMBY

WEATHER

WEATHER

A POPULAR EXPOSITION OF THE NATURE OF WEATHER CHANGES FROM DAY TO DAY

BY THE

Hon. RALPH ABERCROMBY

FELLOW OF THE ROYAL METEOROLOGICAL SOCIETY, LONDON; MEMBER OF THE SCOTTISH METEOROLOGICAL SOCIETY; AND AUTHOR OF "PRINCIPLES OF FORECASTING BY MEANS OF WEATHER CHARTS"

NEW YORK
D. APPLETON AND COMPANY
1887

PREFACE.

The object of this work is the same as that of other volumes of the International Scientific Series, to which it belongs—viz. to place before the general reader a short but clear picture of the modern aspects of the science of which it treats.

With this view, the more elementary parts of the subject—weather science—have been treated of in the first three chapters of the book, while the more difficult questions are reserved for the later portion of the work. Though this method of treatment involves a certain amount of repetition, the author hopes that the work may thus prove acceptable to a large number of readers who would have been deterred from the perusal of a more formal treatise.

This book is not intended to be in any way an encyclopædia of meteorology, or a mere repertory of facts. Our endeavour has been to sketch the great principles of the science as a whole, and to give a clear picture of the general conclusions as to the actual nature of weather to which meteorologists have been led.

Many books have been written on storms and climate, but none on everyday weather. The whole of this work is devoted to weather, in the tropics as well as in the temperate zone.

This volume is not a mere compilation of existing knowledge, for the results of many of the author's original and unpublished researches are included in its pages. Such, for instance, as the explanation of many popular prognostics; the elucidation of the general principles of reading the import of cloud-forms; the classification of those cases in which the motion of the barometer fails to foretell correctly the coming weather; and the character of that kind of rainfall which is not indicated in any way by isobaric maps.

Most of the charts are derived from the publications of various meteorological offices; but almost all the diagrams have been drawn for this work, or have only appeared in some of the author's papers.

Every endeavour has been made to do justice to the discoverers of any new principle, but it has not been considered necessary to give references to all the original authorities in a popular work.

To those who have only known meteorology as a dull branch of statistics, the perusal of these pages may perhaps open a new prospect in science, and a new vision to the mind.

The author wishes specially to acknowledge the assistance which he has received from the Meteorological Office in London, and the United States Signal Office, by the supply of the material contained in some of their various publications; and also the courtesy of the Council of the Royal Meteorological Society of London, in lending him the blocks which have illustrated several of his own papers.

His thanks are also due to Mr. R. Ellery, of Melbourne Observatory, and Mr. W. E. Cooke, of Adelaide Observatory, for information and diagrams to illustrate Australian weather and forecasts; and to Mr. H. F. Blanford, meteorological reporter for the Government of India, for information and material relating to the nature of the monsoons.

CONTENTS.

PART I.—ELEMENTARY.

CHAPTER I.

INTRODUCTORY.

CHAPTER II.

WEATHER-PROGNOSTICS.

CHAPTER III.

CLOUDS AND CLOUD-PROGNOSTICS.

PART II.—ADVANCED.

CHAPTER IV.

ISOBARS.

CHAPTER V.

BAROGRAMS, THERMOGRAMS, METEOGRAMS.

CHAPTER VI.

WIND AND CALM.

CHAPTER VII.

HEAT AND COLD.

CHAPTER VIII.

SQUALLS, THUNDERSTORMS, AND NON-ISOBARIC RAINS.

CHAPTER IX.

PAMPEROS, WHIRLWINDS, AND TORNADOES.

CHAPTER X.

LOCAL VARIATION OF WEATHER.

CHAPTER XI.

DIURNAL VARIATION OF WEATHER.

CHAPTER XII.

ANNUAL AND SECULAR VARIATIONS.

CHAPTER XIII.

TYPES AND SPELLS OF WEATHER.

CHAPTER XIV.

FORECASTING FOR SOLITARY OBSERVERS.

CHAPTER XV.

FORECASTING BY SYNOPTIC CHARTS.

LIST OF ILLUSTRATIONS.

PART I.

ELEMENTARY.

WEATHER.

CHAPTER I.

INTRODUCTORY.

THE earliest records of weather among every nation are to be found in those myths, or popular tales, which, while describing rain, cloud, wind, and other natural phenomena in highly figurative language, refer them to some supernatural or personal agency by way of explanation.

The most interesting thing about these mythical stories is the remarkable fidelity with which they reflect the climate of the country that gave them birth. For example, from the mythologies of Greece and Scandinavia we can almost construct an account of the climate of those two countries by simply translating the figurative phraseology of their legends into the language of modern meteorology.

Many survivals of mystic speech are still found among popular prognostics, and especially in cloud names.

In England and Sweden "Noah's Ark" is still seen in the sky, while in Germany the "Sea-Ship" still turns its head to the wind before rain. In Scotland the "Wind-Dog" and the "Boar's Head" are still the dread of the fisherman, while such names as "Goat's Hair" and "Mare's Tails" recall some of the shaggy monsters of antiquity.

Prognostics.

At a rather later period of intellectual development, the premonitory signs of good or bad weather become formulated into short sayings, or popular prognostics. A large number of these are still current in every part of the world, but their quality and value is very varied. Some represent the astrological attitude of mind, by referring weather changes to the influence of the stars or phases of the moon; others, on the contrary, are very valuable, and, in conjunction with other aids to weather forecasting, prognostics will never be entirely superseded, especially for use on board ship. Till within a very recent period, their science and explanation had hardly advanced since they were first recorded. In many cases the prognostics came true; when they failed, no explanation could be suggested why they did so; neither could any reason be given why the same weather was not always preceded by the same signs. A halo sometimes precedes a storm; why does it not always do so? Why is rain sometimes preceded by a soft sky, and sometimes by hard clouds?

The Barometer.

About one hundred and fifty years ago the barometer was invented. Very soon after that discovery, observation showed that, in a general way, the mercury fell before rain and wind, and rose for finer weather. Also that bad weather was more common when the whole level of the barometer was low, independent of its motion one way or the other, than when the level was high. But as with prognostics, so with these indications, many failures occurred. Sometimes rain would fall with a high or rising barometer, and sometimes there would be a fine day with a very low or falling glass. No reason could be given for these apparent exceptions, and the whole science of barometric readings seemed to be shrouded in mystery.

STATISTICS.

The science of probabilities came into existence about the commencement of this century, and developed the science of statistics. By this method the average readings of meteorological instruments, such as the height of the barometer or thermometer, or the mean direction and force of the wind, at any number of places were calculated, and the results were sometimes plotted on charts so as to show the distribution of mean pressure, temperature, etc., over the world.

By this means a great advance was made. Besides giving a numerical value to many abstract quantities, the plotting of such lines as the isothermals of Dové conclusively showed that many meteorological elements hitherto considered capricious were really controlled by general causes, such as the distribution of land and sea.

Still more fruitful were these charts as the parents of the more modern methods of plotting the readings of the barometer over large areas at a given moment, instead of the mean value for a month or year. We shall refer to the results which have been thus obtained more fully presently. Then by tabulating statistics of the relative frequency of different winds at sea, many ocean voyages —notably those across the "doldrums," or belt of calms near the equator—were materially shortened.

Statistics also of the annual amount of rainfall became of commercial value as bearing on questions of the economic supply of water for large towns, and much valuable information was acquired as to the dependence of mortality on different kinds of weather. Of more purely scientific interest were the variations of pressure, temperature, wind, etc., depending on the time of day, or what are technically known as diurnal variations, which were brought to light by these comparisons.

This branch of the subject is known as "Statistical Meteorology," and has advanced very little since it was first developed by Dové and Kaemtz.

When the attempt was made to apply statistics to weather-changes from day to day, it was found that average results were useless. The mean temperature for any particular day of the year might be 50°, if deduced from the returns of a great many years, but in any particular year it might be as low as 40°, or as high as 60°. The first application of the method was made by the great Napoleon, who requested Laplace to calculate when the cold set in severely over Russia. The latter found that on an average it did not set in hard till January. The emperor made his plans accordingly; a sharp spell of cold came in December, and the army was lost.

It has now been thoroughly recognized that statistics give a numerical representation of climate, but little or none of weather, and that large masses of figures have been accumulated, to which it is difficult to attach any physical significance. The misuse of statistics has done much to bring the science of meteorology into disrepute.

SYNOPTIC CHARTS.

But within the last twenty years a new treatment of weather problems has been introduced, known as the synoptic method, by which the whole aspect of meteorology has been changed. By this method, a chart of a large area of the earth's surface is taken, and after marking on the map the height of the barometer at each place, lines are drawn through all stations at which the barometer marks a particular height. Thus a line would be drawn through all places where the pressure was 30·0 inches, another through all where it was 29·8 inches, and so on at any intervals which were considered necessary. These lines are called "isobars," because they mark out lines of equal pressure. When these charts were first introduced, the estimation of the value of the mean pressure was so great that, instead of drawing lines where pressure was equal at the moment, they were drawn through those places where the pressure was equally distant from the mean of the day for each place. These lines were called "is-abnormals;" that is, equal from the mean. This was, however, soon abandoned, for reasons which will be explained farther on in this work. After the isobars have been put in, lines are usually drawn through all places where the temperature is equal at the moment. These are called "isotherms," or lines of equal temperature. Then arrows to mark the velocity and direction of the wind are inserted; and finally letters, or other symbols, to denote the appearance of the sky, the amount of cloud, or the occurrence of rain or snow. Such a chart is called a "synoptic chart," because it enables the meteorologist to take a general view, as it were, over a large area. Sometimes they are called "synchronous charts," because they are compiled from observations taken at the same moment of time.

When these came to be examined, the following important generalizations were discovered:—

1. That in general the configuration of the isobars assumed one of seven well-defined forms.

2. That, independent of the shape of the isobars, the wind always took a definite direction relative to the trend of those lines, and the position of the nearest area of low pressure.

3. That the velocity of the wind was always nearly proportional to the closeness of the isobars.

4. That the weather—that is to say, the kind of cloud, rain, fog, etc.—at any moment was related to the shape, and not the closeness, of the isobars, some shapes enclosing areas of fine, others of bad, weather.

5. That the regions thus mapped out by isobars were constantly shifting their position, so that changes of weather were caused by the drifting past of these areas of good or bad weather, just as on a small scale rain falls as a squall drives by. The motion of these areas was

found to follow certain laws, so that forecasting weather-changes in advance became possible.

6. That sometimes in the temperate zone, and habitually in the tropics, rain fell without any appreciable change in the isobars, though the wind conformed to the general law of these lines.

Observation also showed that, though the same shapes of isobars appear all over the world, the details of weather within them, and the nature of their motion, are modified by numerous local, diurnal, and annual variations. Hence modern weather science consists in working out for each country the details of the character and motion of the isobars which are usually found over it; just as the geologist finds crumplings and denudation all over the world, and works out the history of the physical appearance of his own scenery by studying the local development of these agencies.

So far the science rests on pure observation—that such and such wind or weather comes with such and such a shape of isobars. But it has been found, still farther, that the seven fundamental shapes of isobars are, as it were, the product of so many various ways in which an atmosphere circulating from the equator to the poles may move. Just as the motion of a river sometimes forms descending eddies or whirlpools, sometimes backwaters in which the water is rising upwards, or yet at other times ripples in which the circulation is very complex, so it now appears that the general movement of the atmosphere from the equator to the pole sometimes breaks up into a rotating and descending movement round that configuration of isobars known as an anticyclone, sometimes into a rotating and ascending movement round that known as a cyclone, or at other times quite in a different way during certain kinds of squalls and thunderstorms.

Isobars, therefore, represent the effect on our barometers of the movements of the air above us, so that by means of isobars we trace the circulation and eddies of the atmosphere.

By carrying the general laws of physics into the conception of a circulating gas, we find that a cold mixed atmosphere of air and vapour descending into a warmer soil would remain clear and bright; while a similar atmosphere rising into cooler strata would condense some of its vapour into rain or cloud. It is by reasoning of this nature that the origin of some of the most beautiful and complex forms of clouds has been discovered.

Following out these lines of research, a new science of meteorology has grown up, which entirely alters the attitude of mind with which we regard weather-changes, and gives rise to an entirely new method of weather-forecasting that far surpasses all previous efforts, and which explains and develops all that was known before.

On the one hand, the new method not only explains why certain prognostics are usually signs of good or bad weather, and the reason why the indications sometimes fail; but also the reason why rain, for instance, is sometimes foretold by one prognostic, and sometimes by a totally different one.

On the other hand, it not only gives a more extended meaning to all the statistics which partially represent the climate of a place, and to the relation of the diurnal to the general changes of weather; but it also enables new inferences to be drawn, which had hitherto been impossible from some observations, and explains why other sets of figures must always remain without any physical significance.

Theoretical Developments.

We may notice here an attempt which has been made by one school of meteorologists to deduce all weather *a priori* from changes in the radiative energy of the sun; that is to say, that from a knowledge of greater or less heat being emitted by the sun, they would treat the consequent alteration of weather as a direct hydrodynamical problem. Given an earth surrounded by fifty miles of damp air, and a sun at varying altitude, and of varying radiative energy, deduce from that all the diverse changes of weather. This is doubtless a very tempting ideal, for there is no doubt that the sun's heat is the prime mover of all atmospheric circulation; but when we have explained what the nature of weather-changes is, we shall see that there is little hope that this method will ever lead to satisfactory results.

Other meteorologists, who lay less stress on the varying power of the sun, have taken up the indications of synoptic charts, and endeavoured to construct a mathematical theory of cyclones and the general circulation of the atmosphere. Ferrel, Mohn, Gulberg, Sprung, and others, have all started with the analysis of the motion of a free mass of air on the earth's surface, first given by Professor Ferrel, and worked out, from that and other general principles, schemes of the nature and propagation of cyclones, and of the general distribution of pressure over the world. Though, as will be seen hereafter, the science of weather-forecasting can never be treated mathematically, still the labours of these writers form a distinct branch of meteorology, and the author regrets that the scope of this work precludes him from giving a chapter which would summarize in a popular manner the results that they have obtained.

This is the deductive portion of meteorology. We shall confine this work entirely to the inductive branch of the science; and, independent of any theoretical considerations, show the observed association of different groups of phenomena, and the generalizations that have been arrived at by observation only.

Plan of the Book.

Though a vast amount of work has been given to synoptic meteorology in all parts of the world, still the results obtained by different investigators remain buried in the scattered transactions of innumerable societies, and no book at present exists which contains a methodical statement of what has been achieved. Many isolated principles have been discovered, but no attempt has been made to lay down the broad principles of the science of weather as a whole. The object of this work is to supply that want by putting before the public a short popular account of all the principal results which have been discovered in recent years by means of synoptic charts, and of their bearing not only in modifying all our views as to the nature of weather at all, but also in explaining all that was previously known. We shall especially endeavour to explain the general principles involved, drawing our illustrations from all countries, so as to show what is general and what is local, and to give a truly International character to this work; while a few examples of British weather will be given in some detail, so as to demonstrate how the minutest weather-changes are subordinate to general laws.

The plan of this book will be as follows. We shall commence with a chapter on popular weather-prognostics, so as to introduce some of the simpler portions of synoptic meteorology. Clouds and cloud-prognostics will form a chapter by themselves, so as to exhibit the great development which has recently been made in the interpretation of their indications. So far, we shall confine our attention to weather of the northern temperate zone only.

This will take up about one-third of the work, and exhaust the more popular portions of the subject. We shall then have to plunge more deeply into the details of isobars, and explain how they are all the products of different forms of atmospheric circulation. From them we shall pass to the consideration of barograms and meteograms generally, and show especially how the changes in the shape of the isobars, as seen on two successive charts, indicate the sequence of weather as observed in any one place. This, which is the fundamental point of all synoptic meteorology, is also unfortunately the most difficult to grasp, and can only be fully realized after considerable practise. Once, however, that its import is fully mastered, the remainder of the work will seem comparatively simple.

We shall then discuss the relation of both the velocity and direction of the wind to isobars, and after that the influence of different shapes of isobars in modifying the distribution of heat and cold from day to day in various parts of the world.

Squalls, thunderstorms, and non-isobaric rains will next engage our attention; and a short chapter on Pamperos, whirlwinds, and Tornados will naturally follow next. We shall then consider the local influence of the configuration of the earth's surface on weather, and devote a whole long chapter to the diurnal phenomena of weather, with special reference to the manner in which they modify the weather that characterizes each shape of isobars.

From this we shall easily be led to comprehend the nature of the annual fluctuations of weather, and of those of a longer period, such as the supposed connection between sunspots and rainfall.

Having thus explained what we may call the components of weather, we shall be ready to understand the nature of sequences or spells of weather, even in the most variable climates. Our illustrations will be drawn chiefly from that portion of the northern hemisphere which lies between the Urals and the Rocky Mountains, but we shall also include some examples from the monsoon districts of India, and from Australia, so as to explain the nature of day to day weather in the tropics and in the southern hemisphere.

When this is done, we shall have completed our description of the nature of weather, and will then turn to the question of forecasting. This falls readily into two distinct problems: 1. To show all that a solitary observer, with a barometer and his eye-observations on clouds and prognostics, can do in the way of forecasting. In this chapter we shall explain fully why the barometer sometimes appears to fail, and also how much the older knowledge can be increased by a knowledge of synoptic charts. The space at our disposal will not, however, permit us to explain the modern developments of the principles of handling ships in hurricanes, which would naturally come in this chapter.

2. To show what a meteorologist can do, seated in a central *bureau*, with telegraphic communication in all directions, and who, after making a synoptic chart, and combining it with every other modern aid, issues telegraphic forecasts to all parts of the country. This is the highest problem of meteorology.

CHAPTER II.

WEATHER-PROGNOSTICS.

Introduction.

The second stage in the history of meteorology, after the mythic phase has been passed, is the collection of numerous observations on the appearance of the sky, the movements of animals, etc., before rain or fine weather into the form of short sayings, which are usually known as popular prognostics. For instance, halos round the sun, or swallows flying low, are known all over the world very frequently to precede rain. On the other hand, a copious deposition of dew, or a white silvery moon, are equally widely known as precursors of fine weather.

One of the earliest collections of prognostics is found in the "Diosemeia" of Aratus, a Greek who flourished in Macedonia and Asia Minor about 270 B.C. The principal interest attached to his work is that many of his prognostics were incorporated by Virgil in his Georgics, and that from them—through the medium of the Latin monks, during the revival of learning in the Middle Ages—a very considerable number have been translated into modern European languages, and are in current use at the present time.

EARLY EXPLANATIONS.

From classic times, down to the commencement of this century, it can hardly be said that this branch of meteorology made any advance. Few, if any, new prognostics had been discovered, and neither their physical explanation nor their meteorological significance had been found out. But about eighty years ago, some physical explanations were given. It was found that the air always contained a certain quantity of uncondensed vapour, and means were invented for measuring this amount accurately. From this, the nature and conditions of the formation of dew were discovered, and also that before many cases of rain the air became more charged with vapour. This latter fact gave the explanation of several rain-prognostics. For instance, when walls sweat, stones grow black, and clouds form on hilltops, rain may be expected almost all the world over.

But even when these reasons had been discovered, the science flagged. A large number of rain-prognostics could not be shown by any means to depend on an increase of moisture, and, as vapour cannot grow in the air, some explanation was needed to account for its variable quantity. And even when, in a general way, the prognostic had been explained, no clue whatever had been found for what we may call the meteorological significance. What was the relation of the damp to the rain? Why did the prognostic sometimes fail? Why are there many ratn-prognostics associated with a tolerably dry air? Why is not all rain preceded by the same set of prognostics? To all these questions no answer could be given. Prognostics had almost fallen into disrepute; they were considered no part of science, and had been supposed to be only suitable for rustics and sailors.

Modern Developments.

So the subject remained till the introduction of synoptic charts. Then it was soon seen that in Temperate regions the broad features of weather depend on the shape of the isobaric lines, and later on it was shown—the author believes, mainly by himself—that nearly all prognostics have a definite place in some shape of isobars, and that all the above questions, formerly insoluble, receive a ready explanation. It has also been demonstrated that prognostics can never be superseded for use on board ship, and that even in the highest developments of weather-forecasting by means of electric telegraph, prognostics often afford most valuable information. But before we attempt to explain how this is done, we must introduce the reader into the elements of synoptic meteorology.

Synoptic Charts.

Synoptic meteorology is that part of the science which deals with the results obtained by constructing synoptic charts. Formerly, all meteorology was deduced from the changes which took place in the instrumental readings at any one place during any interval of time, say one day. For instance, a great deal had been discovered as to the connection between a falling or rising barometer and the accompanying rain or wind. Synoptic charts, on the contrary, are constructed by taking the readings of any instrument (say the barometer), or any observations on the sky or the weather (say where rain is falling, or cloud or blue sky is seen), at a large number of places at the same moment (say 8 a.m. at Greenwich). A map of the area or district from which the observations have been received is then taken, the barometer-readings are marked down over their respective places, and then lines are drawn through all the stations where the pressure is equal. For instance, through all the places where the pressure is 29·9 inches (760 mm.), and again at convenient intervals, generally of about two-tenths of an inch, say 29·7 ins. (755 mm.), 29·5 ins. (750 mm.), and so on. These lines are called isobaric lines, or more shortly isobars—that is, lines of equal atmospheric weight or pressure. This method of showing the distribution of pressure by isobars is exactly analogous to that of marking out hills and valleys by means of contour lines of equal altitude.

Similarly, the places which report rain, cloud, blue sky, etc., are marked with convenient symbols to denote these phenomena. In Great Britain, a system known as Beaufort's weather-notation is exclusively used. It is as follows. This will be useful, as it is employed in all our charts.

BEAUFORT'S NOTATION OF WEATHER.

SYMBOL.

b Blue sky, whether with clear or hazy atmosphere.
c Clouds (detached).
d Drizzling rain.
f Fog.
g Gloomy, very.

SYMBOL.

h Hail.
l Lightning.
m Misty, hazy atmosphere.
o Overcast, the whole sky being covered with an impervious cloud.
p Passing, temporary showers.
q Squally.
r Rain, continued rain.
s Snow.
t Thunder.
u Ugly, threatening appearance of the weather.
v Visibility, whether the sky be cloudy or not.
w Dew.

We should remark here that, though in common parlance the word "weather" is used collectively for the sum of every meteorological element, wind, rain, heat, cold, etc., in this work, and in all synoptic charts, "weather" is used in a more restricted sense to denote whether the actual appearance of the sky is blue, cloudy or otherwise, and whether rain, snow, hail, etc., are falling.

Then arrows are placed over each observing station, with a number of barbs and feathers which roughly indicate the force of the wind. By an international convention, the arrows always fly with the wind; that is to say, they do not face the wind like the pointer of a wind-vane. The scale of force usually adopted is that of Beaufort, which is given opposite. It will be observed that this is a practical scale, based on the amount of canvas a ship can carry. At sea this is certainly a better gauge than any instrumental readings, though there is always a certain disagreement in the estimate of different observers. For land-observations, and those unacquainted with ships, an equivalent of miles per hour and metres per second is given.

Beaufort's Scale of Wind.

Force.			Velocity. Miles per hour.	Velocity. Metres per second.
0. Calm.			3	1·34
1. Light air.	Or just sufficient to give steerage way		8	3·6
2. Light breeze.	Or that in which a well-conditioned man-of-war, with all sail set, and clean full, would go in smooth water, from	1 to 2 knots	13	5·82
3. Gentle breeze.		3 to 4 knots	18	8·1
4. Moderate breeze.		5 to 6 knots	23	10·3
5. Fresh breeze.	Or that to which she could just carry in chase, full and by.	Royals, etc.	28	12·5
6. Strong breeze.		Single-reefed topsails and top-gallant sails	34	15·2
7. Moderate gale.		Double-reefed topsails, jib, etc.	40	17·9
8. Fresh gale.		Triple-reefed topsails, etc.	48	21·5
9. Strong gale.		Close-reefed topsails and courses ...	56	25·0
10. Whole gale.	Or that with which she could scarcely bear close-reefed main-topsail and reefed foresail ...		65	29·0
11. Storm.	Or that which would reduce her to storm-stay sails		75	33·5
12. Hurricane.	Or that which no canvas could withstand		90	40·0

When all this is done, we can see at a glance whether or how wind, rain, cloud, and blue sky are connected with the shape of the isobars. In fact, a synoptic chart gives us, as it were, a bird's-eye view of the weather at the particular moment for which the chart is constructed, over the whole district from which reports have been received. Suppose, now, that after an interval of twenty-four hours another chart is constructed from observations taken over the same area, then we generally find that the shape of the isobars and the position of the areas of high and low pressure have considerably changed, and with them the positions of those areas where the weather is good or bad. For instance, suppose that at 8 a.m. on one morning we find pressure low over Ireland and high over Denmark, with rain over Ireland, cloud over England, and blue sky in Denmark; and that by 8 a.m. on the following day we find that the low-pressure area has advanced to Denmark, and that a new high pressure has formed over Ireland, with rain in Denmark, broken sky in England, and blue sky in Ireland; suppose, too, that the record of the weather, say in London, for those twenty-four hours had been as follows:—cloudy sky, followed by rain, after which the sky broke;—how can an inspection of the two charts help us to explain the weather as observed in London during that day? Our bird's-eye view would show that the rain-area which lay over Ireland in the morning had drifted during the day over England, including London, and covered Denmark by next morning. It would also tell us that the position of the rain was identified with, and moved along with the low pressure. This is the fundamental idea of all synoptic meteorology, but one which can only be thoroughly grasped after a considerable experience in tracing actual cases. It is so different looking at the "ups" and "downs" of the barometer when they are marked on a diagram, and then at any two synoptic charts which refer to the same period, that it is very difficult at first to see any connection at all. In fact, deductions from barograms—as such barometric traces are called—and deductions from synoptic charts are so apparently unconnected that they have hitherto been almost treated as different branches of meteorology. One main feature of this book will be our endeavour to collate these deductions together, and to show how changes in the charts for a large district are simultaneously shown by fluctuation in the instrumental readings at any one place. It must be borne in mind, however, that the whole aim and object of meteorology is to explain weather as it occurs at any place; that is, what successive changes each individual observer will experience. Synoptic charts are only a means to this end.

Relations of Wind and Weather to Isobars.

Such, then, is a synoptic chart. Many thousands have been constructed for all parts of the world, and by comparing them the following important generalizations have been arrived at :—

1. That in general the configuration of the isobars takes one of seven well-defined forms.

2. That, independent of the shape of the isobars, the wind always takes a definite direction relative to the trend of these lines, and the position of the nearest area if low pressure.

3. That the velocity of the wind is always nearly proportional to the closeness of the isobars.

4. That the weather—that is to say, the kind of cloud, rain, fog, etc.—at any moment depends on the shape, and not the closeness, of the isobars, some shapes being associated with good and others with bad weather.

5. That the regions thus mapped out by the isobars were constantly shifting their position, so that changes of weather were caused by the drifting past of these areas of good or bad weather, just as on a small scale rain falls as a squall drives by. The motion of these areas was found to follow certain laws, so that forecasting weather changes in advance became a possibility.

6. That in the temperate zones sometimes, and habitually in the tropics, rain fell without any appreciable change in the isobars, though the wind conformed more regularly to the general law of these lines. This class of rainfall will be called throughout this work "non-isobaric rain."

It will be convenient to take first the broad features of the relation of wind to isobars, which are as follows:—

First as regards direction. The wind in all cases is not exactly parallel to the isobar, but inclined towards the nearest low pressure at an angle of from 30° to 40°. If you stand with your back to the wind, the lowest pressure will always be on your left hand in the northern hemisphere, and on your right in the southern hemisphere. This is what is commonly known as "Buys Ballot's Law."

Then as to velocity. All we need say here is that the velocity is roughly proportional to the closeness of the isobars, and that the measure of the closeness is called the barometric gradient, for in our chapter on wind and calm we will give all necessary details on this branch of the subject.

The upshot of these two principles is, that if you give a meteorologist a chart of the world with the isobars only marked on it, he can put in very approximately the direction and force of the wind all over the globe.

When we have explained the relation of weather to the shapes of isobars, we shall see that he could also write down very nearly the kind of weather which would be experienced everywhere.

The Seven Fundamental Shapes of Isobars.

Then as to the shapes themselves.

In Fig. 1 we give in a diagrammatic form the broad

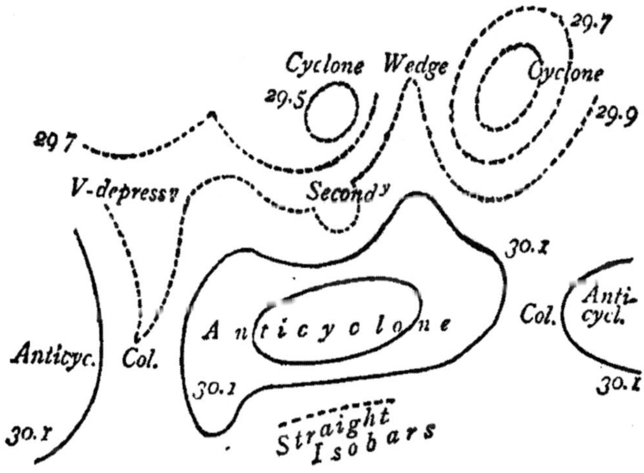

Fig. 1.—The seven fundamental shapes of isobars.

features only of the distribution of pressure over the North Atlantic, Europe, and the eastern portions of the United States on February 27, 1865. Coast-lines are omitted so as not to confuse the eye, so also are lines of latitude and longitude; but the foot-note at the bottom of the figure represents the equator, and the top of the diagram would be on the Arctic Circle. All pressures of and under 29·9 ins. (760 mm.) are shown with dotted lines, so that the eye sees at a glance the broad distribution of high or low pressure. The whole seven fundamental shapes of isobars will be found there.

Looking at the top of the diagram, we see two nearly circular areas of low pressure, round which the isobars are rather closely packed. Such areas, or rather the configurations of isobars which enclose them, are called "cyclones," from a Greek word meaning a circle, because they are nearly circular, and, as we shall see presently, the wind blows nearly in a circle round their centre.

Just south of one of the cyclones, the isobar of 29·9 ins. (760 mm.) forms a small sort of nearly circular loop, enclosing lower pressure; this is called a "secondary cyclone," because it is usually secondary or subsidiary to the primary cyclones above described.

Further to the left the same isobar of 29·9 ins. bends itself into the shape of the letter V, also enclosing low pressure; this is called a "V-shaped depression," or, shortly, a "V."

Between the two cyclones the isobar of 29·9 ins. projects upwards, like a wedge or an inverted letter V., but this time encloses high pressure; this shape of lines is called a "wedge."

Below all these we see an oblong area of high pressure, round which the isobars are very far apart; this is called an "anticyclone," because it is the opposite to a cyclone in everything—wind, weather, pressure, etc.

Between every two anticyclones we find a furrow, neck, or "col" of low pressure analogous to the *col* which forms a pass between two adjacent mountain-peaks.

Lastly, as marked in the lower edge of the diagram, isobars sometimes run straight, so that they do not include any kind of area, but represent a barometric slope analogous to the sloping sides of a long hill.

We may forestall succeeding chapters so far as to say that the cyclones, secondaries, V's, and wedges are usually moving towards the east at the rate of about twenty miles an hour; but that the anticyclones, on the contrary, are usually stationary for days, and sometimes for months together.

We should also note that, though the general principles of prognostics and the broad features of the weather in each of these shapes of isobars are the same all over the world, the minute details which we intend to give now apply to Great Britain and the temperate zones only.

We will now take the five more important shapes separately, and detail the kind of wind and weather which is experienced in different parts of each of them. From this we shall be led to the explanation of the nature of popular prognostics. The account of "V's" and "cols" will be reserved for our chapter on isobars, as no special prognostics are grouped round these two forms.

CYCLONE-PROGNOSTICS.

We will begin with these as they are by far the most important. In Fig. 2 we give a diagram on which we have written in words the kind of weather which would be found in every portion of a typical cyclone; arrows also show the direction of the wind relative to the isobars and to the centre.

First let us look at the isobars. We find that they are oval, and that they are not quite concentric, but the centre of the inner one we will call the centre of the cyclone. Now observe the numbers attached to the isobars; the outer one is 30·0 ins. (762 mm.), the inner one 29·0 ins. (737 mm.). But suppose the outer one was the same,

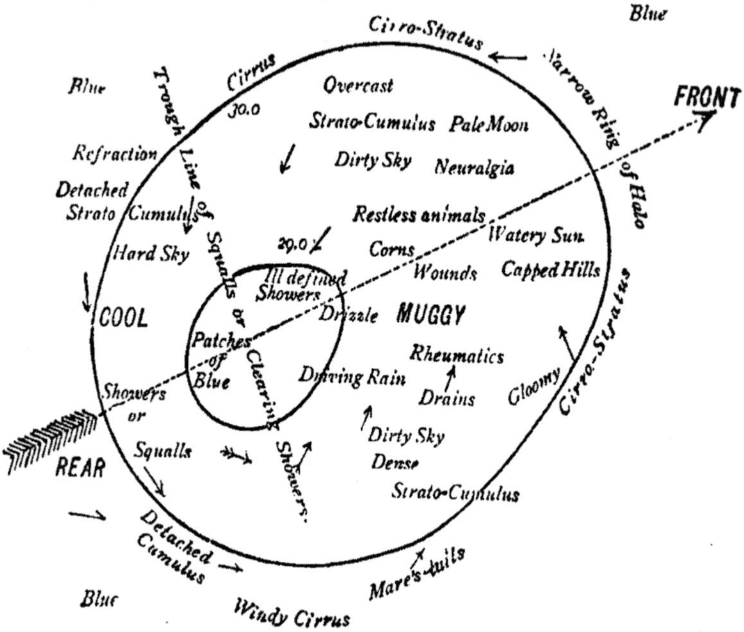

FIG. 2.—Cyclone prognostics.

but the inner one was 29·5 (755 mm.) We should then have two cyclones, differing in nothing but depth; that is, in the closeness of the isobars, or the steepness of the barometric slope. Observation has shown that under these circumstances the general character of the weather and the direction of the wind everywhere would be the same; the only difference would be that the wind would blow a hard gale in the first, and only a moderate breeze in the second case; and that what was a sharp squall in the one would be a quiet shower in the other. This is one of the fundamental principles of synoptic meteorology —that the character of the weather and direction of the wind depend entirely on the shape of the isobars, while the force of the wind and intensity of the character of the weather depend only on the closeness of the isobars.

The difference in the details of the weather in a cyclone, or any other isobaric shape which are due to difference in the steepness of the isobars, is called a difference in the intensity of the weather. Hence, when we speak of a cyclone as being intense, we mean that it has steep isobars somewhere. The word "intensity" will occur very often in these pages, for when we come to talk about the general sequence of weather from day to day, we shall find that *there is no difference between the cyclones which cause storms and those which cause ordinary weather except intensity.* This is another of the fundamental principles of meteorology.

Returning now to our cyclone, the whole of the portion in front of the centre facing the direction towards which it moves is called its front, and the whole of this portion may obviously be divided into a right and left front. The other side of the centre is, of course, the rear of the cyclone. Then, as the whole cyclone moves along its course, it is evident that the barometer will be falling more or less at every portion of the front, and rising more or less everywhere in rear, so that there must be a line of places somewhere across the cyclone, where the barometer has touched its lowest point and is just going to rise.

This line is called the "trough" of the cyclone, because if we look at the barometer-trace at any one place, the "ups" and "downs" suggest the analogy of waves, so that the lowest part of a trace may be called a "trough." Or we may look at the cyclone as a circular eddy, moving in a given direction, and so far presenting some analogy to a wave. Here we are face to face with the primary difficulty of understanding synoptic charts. When we look at any chart of a cyclone which represents the state of things existing at some one moment, there is little to suggest the idea of a trough, because the latter depends on the motion of the cyclone, which cannot be shown on a chart. Perhaps the following illustration may help to explain the nature of the trough. Suppose the cyclone represented the inside of a conical crater, if we walked along the line that marks the path of the centre from the word "FRONT" to the word "REAR" on the diagram, we should pass over the centre of the crater, and be walking downhill all the time till we reached the bottom, and uphill afterwards. But now, if we walked across the crater on any other line parallel to this one, say from the word "*pale*" to the word "COOL," we shall equally walk downhill till we arrive at the point occupied by the letter *q* in the word "squalls." At *q* we should still be on the side of the crater, and some distance from the centre, but after passing *q* we should begin to walk uphill.

When we have once realized the meaning of the trough, we shall never fall into the very common error of thinking that because our barometer has begun to rise, the centre of the cyclone has necessarily passed over us. It is probably only the trough, but we will explain afterwards how we can tell whether it is the centre or not.

So far for the shape and names of the different portions of the cyclone. Now for the wind. A glance at the arrows will show that, broadly speaking, the wind rotates round the centre in a direction opposite to the motion of the hands of a watch. That is to say, that in the extreme front, following the outer isobar, the wind is from the south-east; further round, it is from the east-north-east; still further, from the north-north-west; then from about west; and, finally, from the south-west. Then we note that in front the wind is slightly incurved towards the centre, and therefore blows somewhat across the isobars, while in rear it has little or no incurvature, and blows nearly parallel to the isobars. The velocity or force of the wind will depend on the closeness of the isobars. In the diagram they are much closer set in rear than in front of the cyclone, and therefore the wind is strongest behind the centre.

In our chapter on clouds we shall have to go much more minutely into the nature of wind, both on the surface and in the upper currents; but here we wish to confine our attention as far as possible to the weather and appearance of the sky.

For the same reason, the details of gradients will not be developed till we come to our chapter on wind and calms.

The weather in a cyclone is somewhat complicated. Some characteristic features depend on the position of the trough, and have nothing to do with the centre. For instance, the weather and sky over the whole front of the cyclone—that is, all that lies in front of the trough—is characterized by a muggy, oppressive feel of the air, and a dirty, gloomy sky of a stratiform type, whether it is actually raining or only cloudy. On the other side, the whole of the rear is characterized by a sharp, brisk feel of the air, and a hard, firm sky of cumulus type.

But, on the contrary, other characteristic features are related to the centre, and have little to do with the trough. The rotation of the wind, though slightly modified near the trough, is in the main related to the centre, and the broad features of the weather in a cyclone are—a patch of rain near the centre, a ring of cloud surrounding the rain, and blue sky outside the whole system. The centre of the rain-area is rarely concentric with the isobars. It usually extends further in front than in rear, and more to the south than to the north, but is still primarily related to the centre.

This will be readily seen by reference to the diagram; there the drizzle and driving rain extend some distance to the right front, while almost directly behind the centre patches of blue sky become visible. Thus a cyclone has, as it were, a double symmetry: that is to say, one set of phenomena, such as warmth, cloud character, etc., which are symmetrically disposed in front and rear of the trough; and another set, such as wind and rain, which are symmetrically arranged round the centre. There is reason to believe that what we may call the circular symmetry of a cyclone is due to the rotation of the air, while the properties which are related to the trough are due to the forward motion of the whole system.

As this is a somewhat difficult conception, perhaps

the following analogy may not be out of place. Let us consider the twofold distribution of the population of London. As regards density, we find a comparatively thinly populated district in the centre of London—that is, in the City proper. Round this there is a tolerably symmetrical ring of very densely populated streets, outside of which the population thins away towards the suburbs. But at the same time London is divided into very well-defined halves of comparative poverty and wealth—the east and west ends respectively. This is a far more strongly marked distinction than any which is found between the north and south sides of London, in spite of a river that might have been supposed to make a natural boundary. This distinction into an east and west end is always attributed to the general march of the population westwards. Thus the front and rear of the moving population have a symmetry independent of the density of the population round a centre.

Returning now to the details of weather in a cyclone, we have marked on the diagram the kind of weather and cloud which would be found in different parts of a cyclone. The first thing which will strike us is that the descriptive epithets applied to the sky contain the phraseology of the most familiar prognostics. At the extreme front we see marked "pale moon," "watery sun," which means that in that portion of a cyclone the moon or sun will look pale or watery through a peculiar kind of sky. But all over the world a pale moon and watery sun are known as prognostics of rain. Why are they so? The reason we can now explain. Since a cyclone is usually moving, after the front part where the sky gives a watery look to the sun has passed over the observer, the rainy portion will also have to come over him before he experiences the blue sky on the other side of the cyclone.

Suppose the cyclone stood still for a week, then the observer would see a watery sky for a week, without any rain following. Suppose the cyclone came on so far as to bring him under a watery sky, and then died out or moved in another direction, then, after seeing a watery sky, no rain would fall, but the sky would clear. The prognostic would then be said to fail, but the word is only partially applicable. The watery sky was formed and seen by the observer, because he was in the appropriate portion of the cyclone, and so far the prognostic told its story correctly—viz. that the observer was in the front of the rainy area of a cyclone. The prognostic failed in its ordinary indication because the cyclone did not move on as usual, but died out, and therefore never brought its rainy portion over the observer. This is the commonest source of the so-called failure of a rain-prognostic in Great Britain. The reason why all rain is not preceded by a watery sky is because there are other sources of rain besides a cyclone, which are preceded by a different set of weather-signs. Such is the whole theory of prognostics.

The same reasoning which applies to a watery sky holds good for every other cyclone-prognostic. We shall have explained why any prognostic portends rain when we have shown that the kind of sky or other appearance which forms the prognostic belongs to the front of the rainy portion of a cyclone. Conversely we shall have explained why any prognostic indicates finer weather when we have shown that the kind of sky belongs to the rear of a cyclone. It will be convenient, therefore, to describe the weather in different parts of a cyclone, and the appropriate prognostics together.

First, to take those prognostics which depend on qualities common to the whole front of the cyclone, viz. a falling barometer, increased warmth and damp, with a muggy, uncomfortable feel of the air, and a dirty sky.

From the increasing damp in this part of a cyclone, while the sky generally is pretty clear, cloud forms round and "caps" the tops of hills, which has given rise to numerous local sayings. The reason is that a hill always deflects the air upwards. Usually the cold caused by ascension and consequent expansion is not sufficient to lower the temperature of the air below the dew-point; but when very damp, the same amount of cooling will bring the air below the dew-point, and so produce condensation.

From the same excessive damp the following may be explained:—

"When walls are more than usually damp, rain is expected."

The Zuni Indians in New Mexico say that "When the locks of the Navajos grow damp in the scalp-house, surely it will rain." From this we may assume that scalps are slightly hygroscopic, probably from the salt which they contain.

Also, owing to excessive moisture, clouds appear soft and lowering, and reflect the glare of ironworks and the lights of large towns.

With the gloomy, close, and muggy weather, some people are troubled with rheumatic pains and neuralgia, old wounds and corns are painful, animals and birds are restless, and drains and ditches give out an offensive smell.

A glance at the diagram will show that the barometer falls during the whole of the front of the cyclone. Therefore the explanation of the universally known fact that the barometer generally falls for bad weather is, that both rain and wind are usually associated with the front of a cyclone. When we discuss secondaries, we shall find a kind of rain for which the barometer does not fall; and in our chapter on forecasting for solitary observers we shall explain why it sometimes rains while the barometer is rising, and why there is sometimes fine weather while the mercury is falling.

Now to take prognostics which belong to different portions of the cyclone-front.

By reference to Fig. 2 it will be seen that in the outskirts of the cyclone-front there is a narrow ring of halo-forming sky. Hence the sayings—

"Halos predict a storm (rain and wind, or snow and wind) at no great distance, and the open side of the halo tells the quarter from which it may be expected."

"Mock suns predict a more remote and less certain change of weather."

With regard to the open side of the halo indicating the quarter from which the storm may be expected, it does not appear that this can be used as a prognostic with any certainty. It, however, most probably originated in the fact that halos are usually seen in the south-west or west, when the sun or moon is rather low, the lower portion of the halo being cut off by the gloom on the horizon, and that European storms generally come from those quarters: a heavy bank of cloud will often lie in that direction.

Inside the halo sky comes the denser cloud which gives the pale watery sun and moon. Still nearer the centre we find rain, first in the form of drizzle, then as driving rain. In the left front we find ill-defined showers and a dirty sky.

We have now come to the trough of the cyclone. The line of the trough is often associated with a squall or heavy shower, commonly known as "a clearing shower." This is much more marked in the portion of the trough which lies to the south of the cyclone's centre than on the northern side.

Then we enter the rear of the cyclone. The whole of the rear is characterized by a cool, dry air, with a brisk, exhilarating feel, and a bright sky, with hard cumulus cloud. These features are the exact converse of those we found in the cyclone-front. In the cloud-forms especially we see this difference. All over the front, whether high up or low down, whether as delicate cirrus or heavy gloom, the clouds are of a stratified type. Even under the rain, when we get a peep through a break in the clouds, we find them lying like a more or less thick sheet over the earth. All over the rear, on the contrary, clouds take the rocky form known as cumulus; cirrus is almost unknown in the rear of a cyclone-centre in the temperate zone.

In the exhilarating quality of the air we find the meaning of the proverb—

"Do business with men when the wind is in the north-west."

A north-west wind belongs to the rear of a cyclone, and improves men's tempers, as opposed to the neuralgic and rheumatic sensations in front of a cyclone, which make them cross.

As to the details of the different portions of the rear. Immediately behind the centre small patches of blue sky appear. Further from the centre we find showers or cold squalls; beyond them, hard detached cumulus or strato-cumulus; still farther the sky is blue again.

In the south of the cyclone, near the outskirts, the long wispy clouds known as windy cirrus and "mares' tails" are observed. These indicate wind rather than rain, as they are outside of the rainy portion of the cyclone.

So far we have only described the different kinds of weather which would be experienced at the same moment in different places. We have not said much about the sequence of weather at any one place. A single chart tells little about this, for it does not indicate which way the cyclone is going. To track a cyclone we want another chart about twelve or twenty-four hours later, from which we can see exactly how the cyclone-centre has moved. Then we can follow the sequence of weather for those twelve or twenty-four hours at any place we choose to select.

It must specially be borne in mind that the word "front" is a relative term. In our diagram we have pointed it to the north-east, because that is the direction towards which the majority of British cyclones move. In very rare cases we get a cyclone moving from the south-east. The general circulation of wind then remains about the same, but the characteristic qualities of the different portions of a cyclone are shifted to the new position of the front and rear. For instance, if the cyclone in our diagram was moving towards the north-west, we should have muggy weather and dirty sky with a north-west wind, and bright weather and clear sky with south-west wind. This occurs habitually in the northern Tropics, but very rarely in temperate regions.

But now to take the diagram as it is drawn. We will suppose that the centre has moved along the dotted line, towards the north-east, till it is outside the margin of the figure. What would the sequence of weather be to an observer who was living, say, where the word "*halo*" is written, just below the word "FRONT." This we may get by taking a line across the diagram, parallel to the line which marks the track of the cyclone. This will take us to the word "*detached*," just below the word "REAR." Following the words and symbols, we should find that as the barometer began to fall a halo-forming sky would appear, with the wind coming light from the south-east. Soon the sky would grow lower and denser, into what is known as a "watery" sky, and the wind would begin to veer towards the south and to come in uneasy gusts. Then drizzling rain would set in, the barometer still falling, and the muggy, disagreeable feel of the air would be very noticeable. Later, the wind would begin to rise from the south-west, driving the rain before it, and perhaps attain the force of a gale. After a time one of the gusts would be much harder, with heavier rain than any which had been previously experienced, and with a squall the wind would go round with a jump two or three points of the compass to the west or west-north-west. If we looked at the barometer, we should find that at that moment the mercury had begun to rise; this is the passage of the trough of the cyclone. The wind now blows harder than it had done before, and comes in squalls from the north-west, while the whole aspect of the sky and character of weather are changed. The air is cold and dry, the sky is higher and harder, some patches of blue appear in the heavens, hard rocky cumulus appears on the top of the squalls or showers, and the wind moderates. Gradually showers are replaced by masses of cloud from which no rain descends, and after a time the sky becomes bright and cloudless, while the wind falls to a gentle breeze.

We have endeavoured to show all this in a diagrammatic form in Fig. 3; but observe that, while we read the

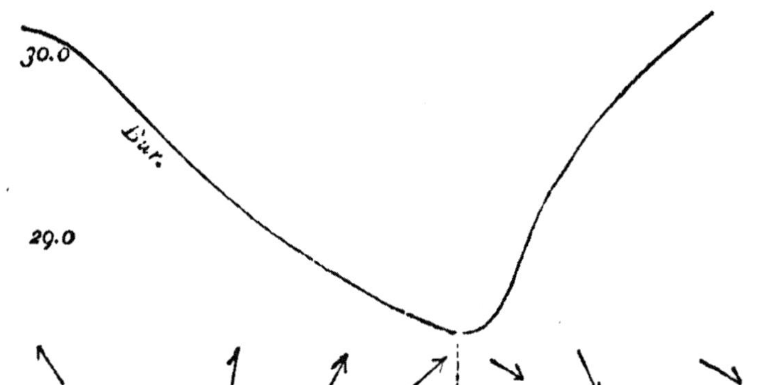

FIG. 3.—Weather sequence in a cyclone.

other diagram from right to left, this one we read in the ordinary manner from left to right. This inversion is obviously necessary, because the cyclone is moving from right to left. The upper line gives the trace which a self-recording barometer would have marked. In front of the cyclone, where the gradients are moderate, the mercury falls slowly; in rear, where the gradients are steep, it rises rapidly. The arrows below, which are supposed to fly with the wind, marks the shift of wind which an observer would experience; and the number of barbs denote the varying force of the wind. The sequence of weather, which is written in words, is identical with the sequence of weather as marked on the plan of cyclone-prognostics.

Wind is said to "veer," or "haul," when it changes in the same direction as the course of the sun; that is, from east, by south, to west, or from west, by north, round to east again. Wind is said, on the contrary, to "back" when it changes against the sun; that is, from east, by north, to west, and from west, by south, to east again. We have seen that the wind veers to an observer situated to the southward of a cyclone-centre. An inspection of Fig. 2 will show that the wind would back from east to north-east, and then through north to north-west, if the observer was situated anywhere north of the cyclone's path. If he was exactly on the path of the centre, the wind would jump round from south-west to north-east without either veering or backing; so that by watching the wind any one can tell what part of a cyclone he is in. In Northern Europe cyclones rarely pass so far to the south as to give the backing sequence. When they do they are almost always soon followed by another cyclone, which passes farther north, and brings fresh bad weather, with another nearly complete shift of wind. Hence the meaning of the following prognostic:—

"When the wind veers against the sun,
Trust it not, for back 'twill run."

Here "veering" is used for shifting generally, and not in its more limited sense of shifting in one particular way.

The explanation which we have just given as to the squall which occurs after the barometer has turned in a cyclone, or at the trough of the cyclone, will show the truth of the following:—

"When rise begins, after low,
Squalls expect and clear blow."

The clear blow refers to the brighter gale which comes from the north-west, as contrasted with the dirty gale which preceded it from the south-west.

If we take a general view of all the weather in a cyclone, we find that we have a large number of prognostics which precede the rain that is associated with wind and a falling barometer. We will now introduce our readers to a kind of rain which is associated with calm and a stationary barometer.

SECONDARY CYCLONES.

A secondary cyclone, or, shortly, a "secondary," is so called both because it has some features in common with a primary cyclone, and because its origin and motion are frequently determined by the path of the primary. It is, however, often found at the edges of anticyclones without any primary, and in many parts of the world it is of frequent occurrence where primaries are almost unknown. The general appearance of that shape of isobars to which we attach the name of secondary will be readily seen by an inspection of Fig. 4. In that diagram the general slope of the isobars is towards the north, but the isobar of 30·1 ins. (765 mm.) is bent into a loop, so as to enclose an area of relatively lower pressure, but not a regular pit of low pressure as in a primary cyclone. In consequence of this the isobaric slope is diminished, as the distance between the two adjacent isobars is increased.

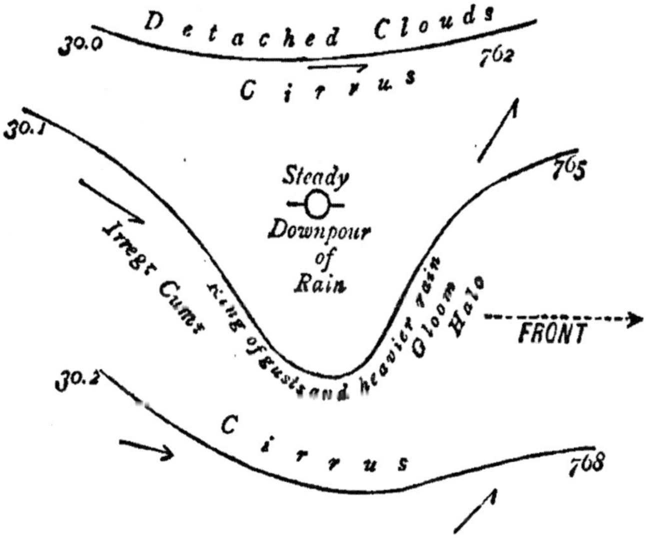

FIG. 4.—Weather in secondary cyclone.

For the same reason, the wind inside the loop is very light, but round the edges of the loop the gradient is increased, and the wind is stronger. This wind is usually in angry, violent gusts, and not in the steady, heavy blow of cyclone-wind.

The motion of the secondary is usually parallel to the path of the primary, and very rarely shows any tendency to revolve round the primary. Hence the term "satellite" depressions, which is sometimes applied to secondaries, seems hardly suitable.

When a secondary is formed at the edge of an anticyclone, the motion is generally very obscure.

Some very striking weather-changes are grouped round this loop of low pressure. In the extreme front we find the thin nebulous cloud which forms halo. Beyond this is a narrow ring of gloomy cirro-stratus; then we find a ring of heavy rain, with gusts, surrounding that half of the secondary where the gradients are steepest. Inside this, in the heart of the secondary, we find a calm, with a steady, heavy downpour of rain. On the rear side of the ring of gusts there is a narrow belt of irregular cumulus, beyond which the sky is blue. On the low-pressure side of the secondary we find cirrus and cirro-stratus, and outside that the cloud appropriate to the primary cyclone. The general course of the wind follows the universal laws of wind and gradient. The arrows in the diagram show that the direction of the wind would be with the slope of gradients which we have drawn there. We may note that the amount of deflection of the wind is much smaller than in primary cyclones. All over the world secondaries are associated with a peculiar class of thunderstorm, but we are unable to say in what particular portion they are localized.

If the primary was moving in the direction of the dotted arrow marked "FRONT" on the diagram, the sequence of weather to a single observer would be as follows. The blue sky would become covered either with a thin nebulous haze, and perhaps halos, or else with dirty cirro-stratus, and the wind would fall light. The clouds would rapidly become black and heavy, and soon, with some angry gusts, heavy rain with big drops would commence suddenly. We saw before that in a primary the rain begins as drizzle. If the barometer is very carefully watched, a very slight rise or fall will occur now; perhaps only one-hundredth of an inch. This gusty rain only lasts a few minutes, when the wind falls, and the rain pours straight down, not quite as heavily as at first. This stage sometimes lasts four or five hours, and is often very puzzling to Englishmen and others who are accustomed to associate rain with a falling, and not with a steady barometer. When the rear edge of the secondary approaches, the rain suddenly becomes much heavier, with more angry gusts. Just at this moment the barometer moves slightly—maybe not more than one-hundredth of an inch. If the general motion of the mercury was upwards when the rain began, this second motion will be upwards also; if downwards, this will be downwards also. Here is another contrast to a primary cyclone, where a fall of the mercury is always followed by a rise. The heavy rain lasts a very short time, when the clouds break quickly into irregular cumulus, and the sky is soon clear again.

We have endeavoured to show this in the annexed diagram (Fig. 5). The upper curve shows the barometric changes to a single observer. In the secondary of which we have drawn the plan in Fig. 4, the general motion of the barometer would be downwards; so in Fig. 5 we find that the small motion of the barometer as the secondary approaches is downwards. Then the mercury remains perfectly steady till the disturbance is just going to pass off, when it takes another small step downwards, and then continues to fall slowly, as before the rain began. Below the barogram, the sequence of weather is shown in a diagrammatic form, so as to show clearly the ring-like character of the rain. The lowest bar, marked overcast, is drawn under the barogram during the whole time that the sky was overcast. The upper shaded bar is drawn double thickness under the portions of the barogram

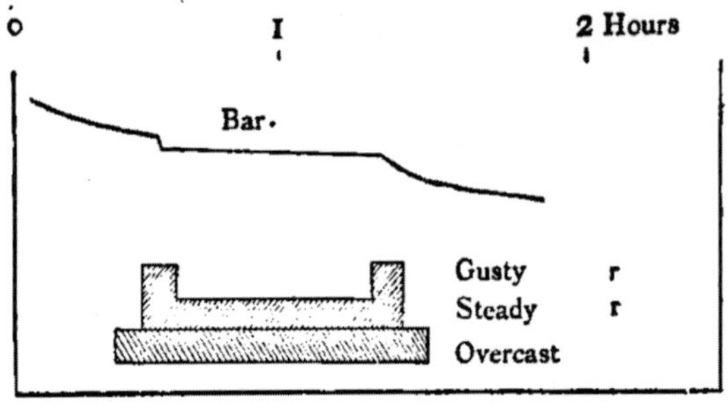

FIG. 5.—Weather sequence in secondary.

during which the heaviest gusty rain was experienced, and single thickness during the time that the steady downpour was observed.

We have already alluded to the idea of intensity in any form of atmospheric circulation; and as simple cloud, moderate, and heavy rain are, as it were, three successive degrees of intensity, the thickness of the bars in the above diagram is proportional to the intensity of the weather in the different portions of a secondary.

There are no special prognostics associated with secondaries. Our object in mentioning the subject here was to explain the nature of another kind of rain than that which is found in primary cyclones. When we come to explain various groups of prognostics, we shall find a number of rain-prognostics which depend on a diminution of barometric pressure. All these indications must obviously be absent before the rain which is produced by a secondary, and we shall now understand why rain often falls without these pressure-prognostics being observed.

Anticyclone-Prognostics.

An anticyclone is an area of high pressure surrounded by nearly circular isobars. These are always a considerable distance apart, and extend over a large area. The pressure is highest in the centre, and gradually diminishes outwards. The air is calm and cold in the central portion, while on the outskirts the wind blows round the centre in the direction of the hands of a watch, not exactly parallel to the isobars, but spirally outwards. Unlike a cyclone, which is commonly in rapid motion, an anticyclone is often stationary for many days together.

Thus in Fig. 6 we see that, while there is a dead calm in the centre, the wind comes from the north on the eastern side, from north-east on the south side, from south-east on the western edge, and from west on the northern edge.

The broad features of the weather in an anticyclone are blue sky, dry cold air, a hot sun, and hazy horizon, with very little wind—in fact, the very antithesis of everything which characterizes a cyclone. As a necessary consequence of this, we find in an anticyclone strongly marked radiation weather, and much diurnal variation. These two last ideas are so important that we must devote a few paragraphs to their explanation and consideration.

Radiation weather is best explained by an example. On a very fine summer day we generally find the valleys full of mist in the early morning. As the sun gains power the vapour rises and evaporates, so that the sky

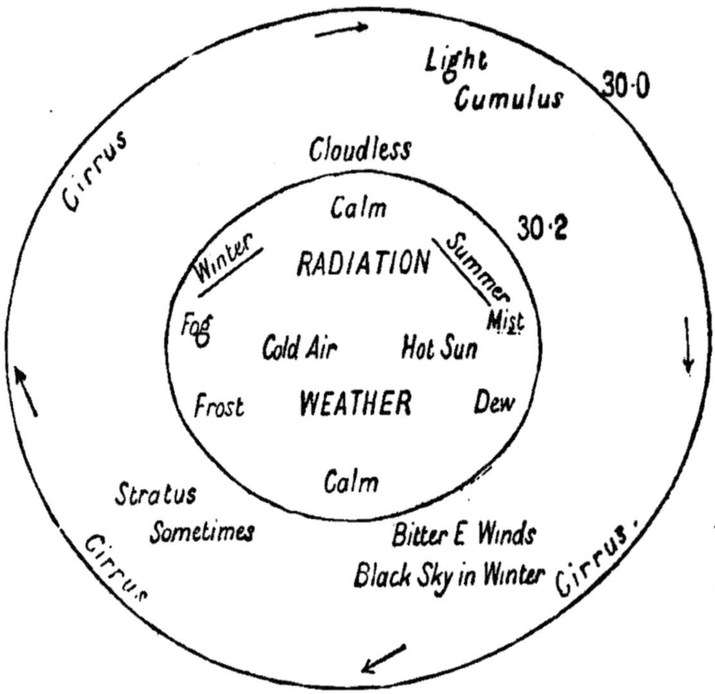

Fig. 6.—Anticyclone prognostics.

becomes cloudless and the sun very hot. After sunset, the air being still and dry, radiation into the cold space which surrounds the earth proceeds rapidly, and soon mist forms again in the hollows, and dew upon the grass.

The sequence of a fine day in winter would be similar in general character, but differ somewhat in details.

Thus, though we have written "fog," "light cumulus," "cloudless," in certain portions of our anticyclone "diagram," these words only describe the prominent daytime weather which most affects us in the various portions of the anticyclone; but the term "RADIATION WEATHER," written in capital letters, denotes accurately the character by night and by day, in summer or in winter.

Moreover, radiation is only a secondary product of an anticyclone. The primary feature is calm; radiation follows as a matter of course, and the weather due to radiation varies enormously in different latitudes, in different seasons, and in different localities in the same country.

The theory of radiation is, that when the air is still the heated surface of the earth radiates into the cold surrounding space, and so the former grows cold enough to condense vapour in the air near the ground, or dew on a suitable surface, such as grass. But when there is more or less wind, each successive layer of air which is in contact with the radiating earth gets removed so rapidly that this condensing process cannot take place, and then no dew or fog is formed. Thus the kind and amount of cloud is variable, but always dependent on radiation.

On the contrary, the particular kind of cloud which forms in any portion of a cyclone is directly and primarily due to the rising current induced by that form of rotating air, and we shall see in a future chapter that the day character of these clouds is not altered during the night.

From the same example of a fine day we can readily pass to the idea of diurnal variation. In the kind of day which we have just described we find a regular sequence from fog to blue, and back again to fog, following the course of the sun—that is, the time day.

But besides the amount of mist, every other meteorological element has a complex series of diurnal changes which depend on the time of day. For instance, both the direction and velocity of the wind have a marked diurnal period, and so has the amount of cloud, the amount of vapour, and every other component of weather. These, and many others, will form the subject of a subsequent long and rather difficult chapter. All that we have to note here is the important principle of meteorology —that the primary character of all weather is given by the shape of the isobars, whether cyclonic, anticyclonic, or otherwise; that a complex series of diurnal changes are superimposed on this, which modify, but do not alter the intrinsic quality; and that the resulting weather is the sum of the two together. Thus in a cyclone the changes of weather due to its motion are so marked and so strong that diurnal changes are often entirely obliterated; while in a calm anticyclone, where there is no motion, the uniform character of the general weather allows full play to radiation, and the diurnal changes are very prominent.

Throughout this work we shall call the character and changes of weather which are due to the shapes of the isobars, the general character and general changes, because they are caused by alterations in the general distribution of pressure over a large portion of the earth's surface. On the contrary, changes which are due to the time of day, the season of the year, or to any local peculiarity, we shall call diurnal, seasonal, or local variations of the general character. The first are really changes, the second only variations. The reason why many prognostics which are due to radiation and diurnal causes are signs of settled fine weather, is because in a country like England they can only occur in an anticyclone. An anticyclone means settled fine weather, not only because the weather at any moment in it is fine, but because it is usually stationary, and so there is nothing to change the existing conditions. All anticyclone prognostics fail when the anticyclone breaks up suddenly, or in the not very common case when it moves onward along a definite path.

We will now give a few prognostics due to the variations of an anticyclone in some detail.

The sky being generally clear and the air calm, the temperature is high in the day and low at night. In summer brilliant sunshine prevails during the day, and at night there is a heavy dew, and, in low-lying places mist.

"Heavy dews in hot weather indicate a continuance of fair weather, and no dew after a hot day foretells rain."

"If mists rise in low ground and soon vanish, expect fair weather."

Fine, bright, genial weather raises the spirits and exerts an enlivening influence not only on human beings, but also on animals, birds, insects, etc. Hence the sayings—

"When sea-birds fly out early and far to seaward, moderate winds and fair weather may be expected."

"If rooks go far abroad, it will be fine."

"Cranes soaring aloft and quietly in the air foreshadows fair weather."

"If kites fly high, fine weather is at hand."

"Bats or field-mice coming out of their holes quickly after sunset and sporting themselves in the open air, premonstrates fair and calm weather."

"Chickweed expands its leaves boldly and fully when fine weather is to follow."

These are merely samples of innumerable similar prognostics in all parts of the world.

In winter frost is generally prevalent in the central area of an anticyclone, accompanied frequently by fog, which is most dense in the neighbourhood of large towns. This is all due to the radiation of calm weather.

"White mist in winter indicates frost."

The wind is usually very light in force.

"It is said to be a sign of continued good weather when the wind so changes during the day as to follow the sun."

This "veering with the sun," as it is called, is the ordinary diurnal variation of the wind, which in England is only very obvious with the shallow gradients of an anticyclone. At seaside places in summer very often "the wind is in by day and out by night," which is the equivalent of the land and sea breezes of the tropics. Like the preceding prognostic, it is only in anticyclones that local currents of air, probably due to unequal heating of sea and land, can override the general circulation of the atmosphere in this country.

Sometimes in winter, on the southern side of the anticyclone, bitter east winds with a black-looking sky will prevail for several days together, when it may truly be said—

> "When the wind is in the east,
> It is neither good for man nor beast."

This class of anticyclone prognostics hold good as long as the anticyclone remains stationary. Occasionally the anticyclone moves on, and is replaced by some other form of isobars; but far more frequently the anticyclone breaks up—that is to say, it disappears without moving on, and is replaced by a cyclone or some other type of isobars.

Wedge Shaped Isobars.

We have already defined wedge-shaped isobars as a projecting area of high pressure moving along between two cyclones. This wedge may point in any direction, but in practice by far the most frequently to the north. We have therefore selected such a one for the diagram of the wind and weather in an ideal wedge, which we give in Fig. 7. There the highest pressure is at the bottom of the diagram, while the wedge-shaped isobars project towards the north. On the right hand we see the rear of a retreating cyclone; on the left, the front of an advancing depression. As these two cyclones move forward, the wedge goes on between them, so that there must always be a line of stations where, after the barometer has risen owing to the onward passage of the first cyclone, the mercury has just begun to fall, owing to the advance of the second depression. This line is called the crest of the wedge, and is marked by a dotted line in the diagram.

The wind blows round the wedge in accordance with the universal law of gradients. Thus on the east side of the wedge the wind is from north-west; in the centre it is calm; and on the west side, from south-west to south-

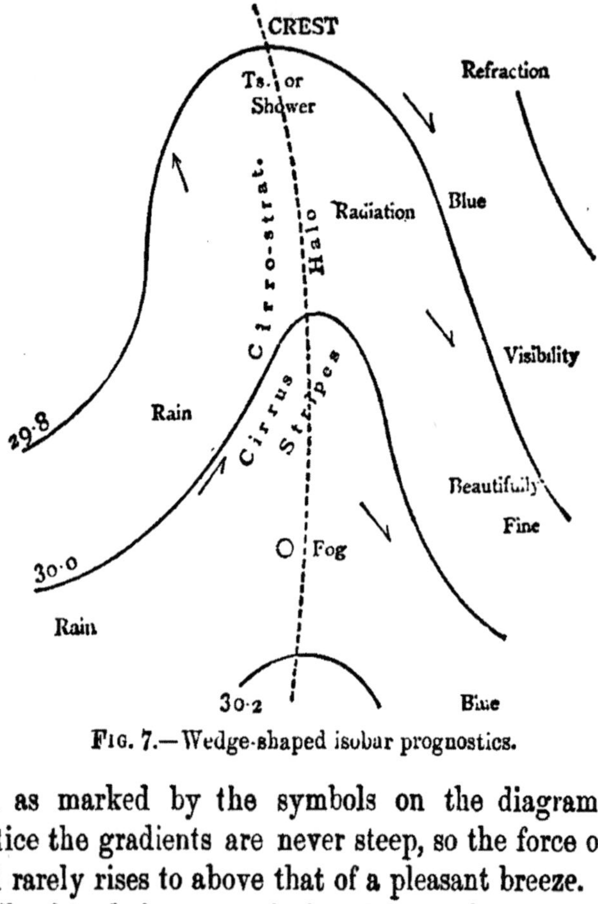

Fig. 7.—Wedge-shaped isobar prognostics.

east, as marked by the symbols on the diagram. In practice the gradients are never steep, so the force of the wind rarely rises to above that of a pleasant breeze.

The broad features of the cloud and weather in a wedge are written across the diagram (Fig. 7). In front we find blue sky, with beautifully fine weather, refraction, and that unusual clearness of the atmosphere known as "visibility." Nearer the calm under the crest we come to radiation weather, with fog; and then to halo-bearing sky just in front of the crest, and stripes of cirrus-cloud. A thunderstorm or heavy shower is often experienced at the top of the wedge. In rear of the crest the sky becomes covered with cirro-stratus cloud, and further in rear we find the rain of the approaching cyclone.

Here, as in cyclones, we see the striking fact that the words applied to describe the weather contain the phraseology of many familiar prognostics, such as those connected with visibility, halos, or the stripes of cirrus which form the cloud popularly known as "Noah's Ark."

If we remember how we took a sort of section across a cyclone, and so found the sequence of weather at any station, we shall readily understand that we have only to read this diagram (Fig. 7) from right to left to get the sequence of weather during the passage of a wedge. Thus we should have a beautifully fine day, with a north-west wind and a rising barometer, with a hot sun by day, and a cold night with radiation according to the season of the year. Then, while the barometer was still rising, the blue sky would assume that peculiar nebulous whiteness which forms halos, and stripes of cirrus would appear in places. Soon the barometer would begin to fall, the sky to grow denser and overcast, and before long the drizzling rain of the new cyclone would begin to fall, the wind having previously backed to the south-west.

We now see the meaning of the halo and cirrus-stripes being marked in the diagram partly in front of the crest of the wedge, viz. that in a wedge the sky shows the approach of a new cyclone before the barometer at a single station has ceased to rise. This is very interesting, as it is the first opportunity we have had of explaining why the barometer sometimes appears to fail, and rises, as in this case, while bad weather is manifestly approaching. It also shows the great additional power of forecasting which the use of synoptic charts gives to a meteorologist, seated in a central office, with abundant telegraphic communication. Suppose any morning that a forecaster found his isobars were wedge-shaped. He could then telegraph to the eastern districts of his territory that the fine weather would not last, though they with their rising mercury might have thought the contrary.

It may be remarked here that all cyclones are not preceded by a wedge, but only those which roll, as it were, along the northern edge of large stationary anticyclones.

We can now explain in detail the prognostics that are marked on the diagram, and several others for which there was no room. Any appearance of the sky which characterizes the front of a wedge will be a sign of rain, because there is always rain in rear of that shape of isobars. These prognostics of wet which are associated with fine, dry weather are particularly interesting, because they are the very opposite of the rain-prognostics in a cyclone, which are associated with increasing damp and a dirty sky.

It used to be thought that every prognostic of rain would be explained by showing that the appearance was due to an increase of vapour in the air; here we find that the prognostic can only be explained on the supposition that many cyclones develop an area of calm and clear sky in front of, and as a portion of, themselves. It is a crude idea of meteorology to think that all rain-prevision depends on hygrometry.

Returning now to Fig. 7, we see that in the rear of the retreating cyclone the air is dry and the weather beautifully fine—of the sort of which we would say that it was "too fine to last;" or, if it lasted a whole day, we should talk of a "pet day."

During the day the sun is burning hot.

"When the sun burns more than usual, rain may be expected."

During the night white frost is formed, owing to calm radiation.

"A white frost never lasts more than three days; a long frost is a black frost."

"Frost suddenly following heavy rain, seldom lasts long."

As the day advances, after a white frost, the air becomes dull from the influence of the on-coming depression. Whence the saying—

"When the frost gets into the air, it will rain."

During the very fine weather on the east side of a wedge-shaped area there is often great visibility, with a cloudless sky.

"The further the sight, the nearer the rain."

This is one kind of visibility; there is another class that is associated with a hard, overcast sky, as we shall explain under "straight isobars," and also visibility in the tropics, which depends on causes which cannot be explained here. From the same transparency of the atmosphere, the "ashy" light of the dark portion of a new moon is very strong.

"If the old moon embraces the new moon, stormy weather is foreboded." Great confidence is placed in this old prognostic.

> "I saw the new moon, late yestreen,
> With the old moon in her arm,
> And I fear, I fear, my master dear,
> We shall have a deadly storm."

At the extreme north-west edge of a cyclone there is often a particular kind of "refraction"—a well-known sign of rain. This seems to be due to the cold air in the rear of a cyclone being much below the temperature of the sea. If so it is a sign of rain, for the reason that one cyclone is usually soon followed by another. There is another kind of refraction caused by a cool south-east wind in an anticyclone blowing over a heated sea, which is usually a sign of fine weather. This is a good illustration of how the same prognostic may portend either good or bad weather, according to its surroundings.

If the cyclone in front of the wedge has produced a north-west gale, it is not improbable that the on-coming one may begin with a south-west gale. Hence the significance of the well-known nautical saying in the Atlantic—

"A nor'-wester is not long in debt to a sou'-wester."

In the cyclone and secondary we have found rain of different kinds; so in the anticyclone and wedge we have found fine weather of different kinds. Anticyclone fine weather is almost always hazy, and is settled weather, because the anticyclone itself is usually stationary. Wedge fine weather is always clear, and is only temporary because the wedge is never stationary. Hence we see that when we talk of rain and fine weather, it is often necessary to say what kind of rain and what kind of fine weather; and we find, moreover, that a knowledge of the kind of isobars enables us to define the kind of weather. In many discussions on climate and statistical meteorology, we find that terrible confusion is caused by mixing up together all kinds of good and bad weather.

The prognostics which are associated with a wedge are almost less liable to failure than those which accompany other shapes of isobars. When they do fail, it is usually from a sudden break-up of all the existing distribution of pressure.

Straight Isobars.

Straight isobars are so called because the isobars have no curvature. The trend of the lines may be in any direction, and so may their slope. For instance, the lines may lie east and west, but the slope may be either towards the north or towards the south. In our general diagram (Fig. 1) of all the fundamental shapes of isobars, we drew some straight isobars sloping to the south. In temperate regions this slope is uncommon, while a slope to the north or north-west is very common. We have therefore selected an instance of a northerly slope for our diagram of straight isobars in Fig. 8, and, as before, have written in words the kind of sky and weather which we find in different parts of the slope. In all the other shapes of isobars which we have hitherto described, the lines enclose an area of high or low pressure, while in straight isobars the lines only mark the position of what may be called a barometric slope.

On turning to Fig. 8, it will be seen that while the pressure is high to the south, it is generally low to the

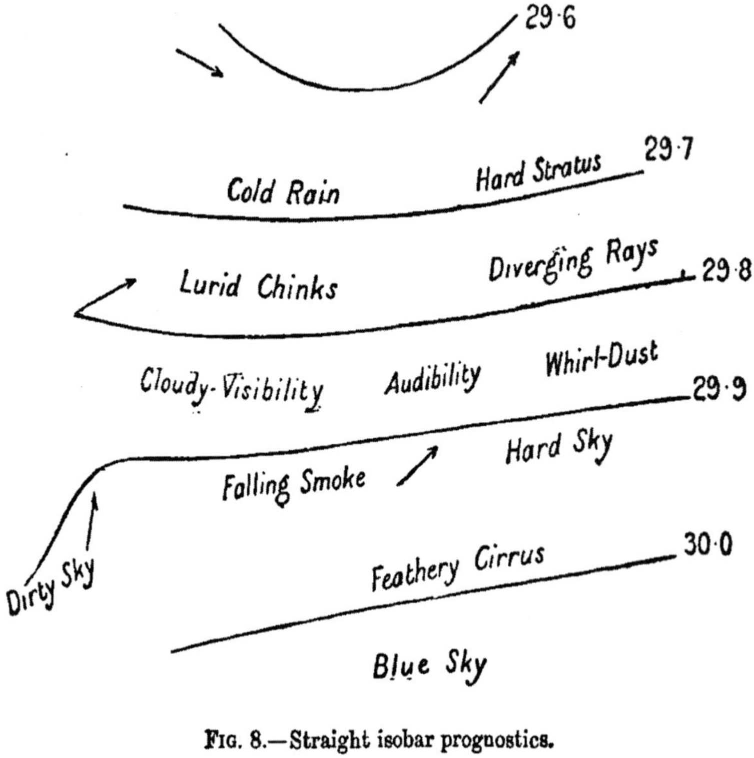

FIG. 8.—Straight isobar prognostics.

north, without any definite cyclonic system, and that the isobars run straight nearly east and west, with a slope towards the north. The wind is from the south-west or west, and usually strong and gusty, but short of a gale. On the high-pressure side the sky is blue; then as we approach the low-pressure, feathery cirrus, or some form of windy sky, makes its appearance, while a blustery wind whirls the dust or blows the soot down. The falling of soot refers to blacks falling out-of-doors and coming into windows from being blown about. Sometimes, in very damp weather, soot seems to fall from condensation of vapour on itself, and at other times masses of soot fall down a chimney from the action of hail or very heavy rain.

Getting still nearer the low-pressure, the sky is found to be gathering into hard strato-cumulus, at first with chinks between its masses, through which divergent rays stream down under the sun, which is spoken of as "the sun drawing water." Sometimes, especially in winter, these rays are lurid, and the appearance of the sky is then very striking. This prognostic is common all over Northern Europe, and in Denmark takes the form of "Locke is drawing water." Loki is a well-known demi-god in the Scandinavian Eddas, so that we have here a direct survival of mythic speech. This hard strato-cumulus is especially characteristic of straight isobars in Great Britain.

At the same time there is often great "visibility," with a hard, overcast sky and moderately dry air, in which the cloud seems to play the part of a sunshade, for as soon as the sun comes out the clearness of distant objects diminishes. This visibility must not be confounded with the visibility already described with a cloudless sky, which occurs with wedge-shaped isobars.

Simultaneously we often find "audibility." This distinctness of distant sounds must be carefully distinguished from sounds which are not usually heard, being brought up by the wind coming from a rainy quarter. For instance, the whistle of a railway-train to the south of a house will not be usually heard with the normal south-west wind of Great Britain; but when the wind backs to the south in front of a depression, then the noise will be heard; and though this will be a good prognostic, still, it is not true audibility.

When the gradients are very steep, a little rain sometimes falls with straight isobars, generally in light showers, with a hard sky.

Though, as a matter of convenience, we have described the sequence of weather as we proceed from the high to the low pressure, it must be clearly understood that it does not represent the sequence of weather to a single observer, but rather what the weather will be simultaneously in different parts of the country; for instance, that if there is cirrus in London, there may perhaps be a lurid sky in Edinburgh.

But now audibility, visibility, whirling dust, and lurid chinks with divergent rays are well-known signs of rain almost all over the world, so we have to explain why the appearance of the sky in straight isobars is a sign of rain. It is found by experience that straight isobars are never persistent, and that, practically, the district which they cover one day will be traversed by a cyclone the next day. It does not follow that the cyclone is necessarily in existence when we observe the straight isobars; but, from the nature of weather-changes, straight isobars seem to be an intermediate form of atmospheric circulation which precedes the formation of a cyclone.

We cannot, therefore, draw a section across straight isobars and say that it will give the sequence of weather at any place, for we are not dealing with a moving form of pressure, but with a transitional state of things which cannot last long. The chief interest of these rain prognostics lies in the contrast which they present to those associated with a cyclone. While those in a cyclone are accompanied by an almost ominous calm and a dirty, murky sky, these are associated with a hard sky and blustery wind, of which it would be ordinarily remarked "that the wind keeps down the rain," or, "that when the wind falls, it will rain." While, also, the prognostics which precede cyclone-rain hold good for the reason that they are seen in front of the rainy portion of such a depression, those associated with straight isobars hold good because, though there is little rain actually with them, the area which they cover to-day will probably be covered by a cyclone to-morrow—the conditions being favourable for the passage of depressions. Another point of contrast lies in the comparative dryness of the air in straight isobars, as compared with the excessive amount of moisture which precedes cyclones. The same remarks apply to these as to the fine-weather prognostics associated with wedge-shaped isobars.

All the prognostics we have discussed under this heading fail when the straight isobars are formed during a general rearrangement of the whole distribution of pressure over the northern hemisphere, because a cyclone may not then traverse the district where the well-known signs of rain had been observed.

The other fundamental forms of isobars—V-shaped

depressions and cols—are not associated with any distinctive prognostics, so we will defer our consideration of these shapes till a subsequent chapter.

GENERAL REMARKS.

We are now in a position to take a general survey of the whole principle of prognostics, and to answer the questions which, we mentioned at the commencement of the chapter, were formerly considered insoluble. What is the place of prognostics in meteorology, and how has modern research developed their utility? Why do prognostics sometimes fail? Why are not all prognostics associated with increasing damp? Why is rain or fine weather not always preceded by the same prognostic?

The details which we have already given abundantly show that every portion of every shape of isobars has a characteristic weather and look of sky, and that prognostics simply describe these appearances.

Theoretically, then, when the isobars are well defined, we ought to be able to write down the prognostics which might be visible everywhere, but practically we cannot do so completely; and also, theoretically, all that any prognostic does is to enable a solitary observer to identify his position in any kind of atmospheric circulation. Thus the associates of the front of a cyclone or secondary are signs of bad weather; while those of the rear of a cyclone, or of any portion of an anticyclone, are signs of fine weather. The word "front" implies not only the idea of motion, but also of the direction of that motion. But here comes in the reason why prognostics can never develop the science of forecasting much further than at present. From the nature of cyclone-motion, as will be abundantly illustrated in a future chapter on Types of Weather, these depressions have a way of advancing so far in a certain direction, and then of either changing their front or else of dying out altogether. No prognostic can give any clue to the probability of either of these changes. On the other hand, a forecaster in a central office, with synoptic charts, can often tell when a cyclone is going to be arrested or deflected from its previous course, and this branch of the science is undoubtedly capable of very great extension.

Then the question may very naturally be asked, How far the introduction of synoptic charts has developed our knowledge of prognostics? So far as explaining their true nature, the advance has been very great; but so far as increasing their practical utility, the progress has been much less. The new explanations have already been given. The principal improvement in reading the indications of prognostics is rather that the observer can distinguish between the different kinds of rain and fine weather, and so give greater precision to his previsions, than that he can alter the reputed value of any weather-saying. For instance, if he sees a halo, with a falling barometer and increasing wind, he knows that he is in for the whole sequence of a cyclonic storm; whereas, if he sees a halo, with a steady barometer and a few angry gusts, he knows that he need only expect heavy rain without any great wind.

We have already mentioned why prognostics sometimes fail, either from the alteration of a cyclone's front, the breaking up of an anticyclone, and other similar reasons. For those who are unfamiliar with the nature of weather-changes, an actual example will be more acceptable than a general statement. We will select an instance of the failure of a halo-prognostic. Let us look at the chart in Fig. 9 for February 5, 1883, at 8 a.m. At Folkestone, near Dover, in England (marked F in the

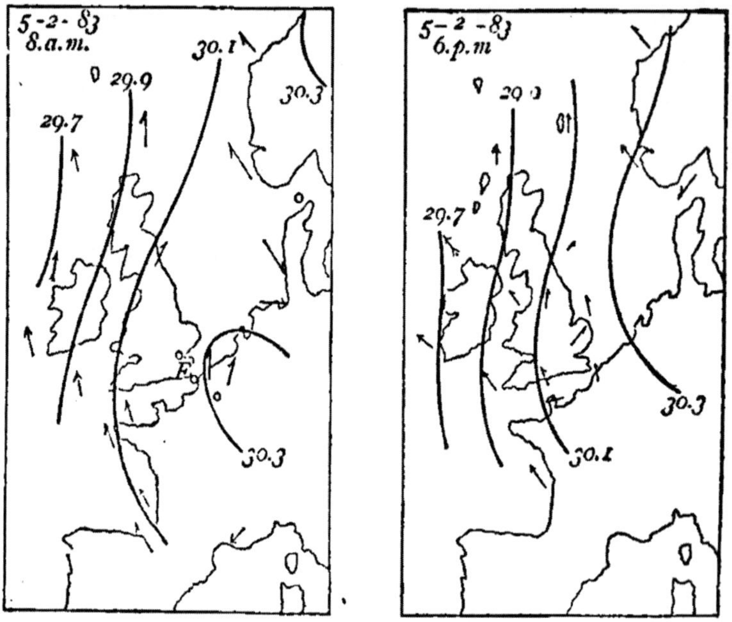

FIGS. 9 and 10.—Illustrating the failure of the prognostic that halo indicates wind or rain.

diagram), a halo was visible off and on from 9.30 a.m. to 4 p.m., and that was due to the front of the cyclone which is seen lying off the north-west coasts of Great Britain. The whole of that day and night, as well as the succeeding day, was very fine, so that the prognostic might seem to have failed. Before the days of synoptic charts this is all that we could have said about the matter, but now we can explain the reason why.

In the first chart (Fig. 9) we see an anticyclone marked by the isobar of 30·3 ins. over France, and the extreme edge of another over part of Norway. The col between them covers Denmark and the North Sea. The extreme rear of a cyclone is found near Copenhagen, while on the north-west of Ireland a new cyclone of enormous diameter is approaching. The halo-forming portion of this last has almost reached Dover, while the rainy portion is already causing precipitation over Ireland. If the cyclone continued its course, in due time the rain-area would reach England; but suppose the cyclone stood still, or some rearrangement of pressure arrested its onward progress, what would happen then? Why, the halo-prognostic at Dover would fail; that is, would not be followed by rain and wind, as is usually the case. Now, this is exactly what happened. If we look at Fig. 10, which gives the synoptic conditions after an interval of ten hours—at 6 p.m. the same day—we see that the two small anticyclones have coalesced into a single large one, which lies over Sweden and North Germany, while the Irish cyclone has disappeared, instead of moving as usual to the north-east; and that a barometric slope, with nearly straight isobars, covers Great Britain and France. In consequence of these changes, the weather remained fine all day near Dover, and so the prognostic appeared to fail.

We will explain, in another chapter on Forecasting by Synoptic Charts, why a forecaster in a central bureau could not have announced with certainty that there would have been no rain in that portion of England; and also when and how prognostics sometimes assist him in foretelling rain which he would hardly have expected from the mere inspection of the isobars.

These charts also give an idea of the extreme rapidity of meteorological changes in a climate like that of Western Europe. The short space of ten hours has been sufficient not only to alter very materially the general distribution of pressure, but also to form new configurations of isobars. These charts also show that changes of weather are not only caused by the passage of well-defined shapes of isobars, such as cyclones, in the manner which we have just described, but also by the readjustment of pressure-distribution and the formation of new isobaric shapes over the area under observation. This last conception is of fundamental importance.

Then we come to the question why all rain-prognostics are not associated with increasing or excessive damp. The answer to this is that there are different kinds of rain, such as the rain in front of a primary cyclone, which is associated with great damp; and the light showers of straight isobars, which are associated with a rather dry air. Also that some rain-prognostics, such as those associated with the much too fine weather in front of a wedge, owe their value to the fact that a wedge precedes a cyclone, though the air in itself is tolerably dry.

A similar train of argument applies to the question why rain and fine weather are not always preceded by the same prognostics. We have just mentioned two different sorts of rain, and, as regards fine weather, we need only mention that in a like manner there are many kinds of fine weather. For instance, the formation of small blue patches in an otherwise overcast sky in rear of a cyclone foretells one kind of fine weather, while the radiation phenomena of an anticyclone also indicate fine weather, but of a totally different sort.

We have shown why the value of the indications which prognostics afford can never be materially improved, but at the same time no advance in synoptic meteorology will ever supersede the use of prognostics. Our own researches on hurricanes in the tropics have proved that there, as in Europe, unusual colouration of the sky at sunrise and sunset apparently often precedes the formation of any notable barometric depression; so that sometimes the indications of prognostics are ahead of those of any other system of forecasting. In isolated stations, and on board ship especially, an observer must always rely to a great extent on his own eyes to gather information from the aspect of the sky as well as from the readings of his own barometer. We shall, in fact, devote a whole chapter at the end of this book to the consideration of the problem of how much weather-forecasting a solitary observer can do for himself. In the following chapter we shall consider a great many prognostics connected with clouds which could not well be associated with definite shapes of isobars.

CHAPTER III.

CLOUDS AND CLOUD-PROGNOSTICS.

In this chapter we propose to discuss the nature of clouds by first explaining their origin and the varying conditions under which they are formed.

This will lead to a classification of their different shapes and forms, and give us a certain insight into the varying velocity and direction of the upper currents of the atmosphere. But when we come to the more modern developments of cloud-knowledge, we shall have to consider the relation of clouds to the great areas of high and low pressure, which we have already described as cyclones and anticyclones.

Some of this we have already seen in our chapter on prognostics, where we showed that different kinds of cloud are characteristic of different portions of cyclones, etc. But in this chapter we will explain how, from a study of cloud-motion in the upper strata, we are enabled to discover much about the real nature of the circulation of the air in the different shapes of isobars. In the course of our remarks, we shall explain incidentally both the meaning and value of the older cloud-lore, and also the great development in the science of forecasting by means of clouds which has been made by recent researches.

Nomenclature of Clouds.

Unfortunately, in approaching the subject of cloud-nomenclature, we come to one of the most unsatisfactory branches of meteorology. Though the words of Howard's nomenclature are universally employed, the same word is by no means always employed for the same kind of cloud; and for this reason, though the words we shall employ to designate clouds are those which are used by many, we shall be very careful to describe exactly the kind of cloud we mean by any particular name.

For practical purposes clouds are divided into four different classes, according to their most obvious differences of shape; but these classes are only as a matter of convenience, for in nature they all run into each other by imperceptible gradations. The forms are—

1. *Cumulus.* All cloud which has a rocky or lumpy look is either pure cumulus or must contain the word *cumulo* in combination with some other name.

2. *Stratus.* All cloud which lies as a thin flat sheet must either be pure stratus or contain the word *strato* in combination.

3. *Cirrus.* All cloud which has a wispy, feathery, or curly look must either be pure cirrus or must contain the word *cirro* in combination.

4. *Nimbus.* Any cloud from which rain is falling is nimbus in some form.

It must be fully understood that these names and their derivatives, which we shall give presently, do not by any means exhaust all the varieties of clouds which very experienced observers can detect and classify. All that we propose to give here are the broad distinctions which anybody can understand, for all meteorologists who have to deal with corps of observers are agreed that eight or ten names are as many as can practically be employed.

We will first explain the simple forms of these clouds, and then the more complicated combinations, such as cirro-stratus, cirro-cumulus, cumulo-stratus, etc. But besides giving a broad classification to the leading kinds of cloud, these terms also give a rough relative scale of altitude. Thus in practice stratus and cumulus are usually the lowest, the composites the middle, and cirrus the highest layer of cloud; but no absolute level can be assigned to each stratum at any season, or in any country. For instance, cumulus may be as low down as 2000 feet, and cirrus as about 12,000 feet; and, on the other hand, cumulus may be formed up to at least 25,000 feet, and cirrus probably up to at least 50,000 feet; but true cirrus can never be formed under cumulus, whatever the relative latitudes may be.

These relative heights also partially determine the nomenclature. If a cloud is very high up, we have to add the word *cirro*, to indicate altitude, to the word which denotes the form only; while *cumulo* would suggest a lower level. The first word before a compound name gives the idea of relative altitude: thus cirro-cumulus is higher than cumulo-cirrus.

Cumulus.

Pure cumulus may be described as convex or conical heaps increasing upwards from a flat horizontal base, as in Fig. 11, *a*. It is undoubtedly formed by the condensation of the summit of an ascensional column of vapour-laden air, as shown by the dotted lines. When this is cooled, either by rising into a colder stratum than that from which it started, or by expansion, the water-vapour

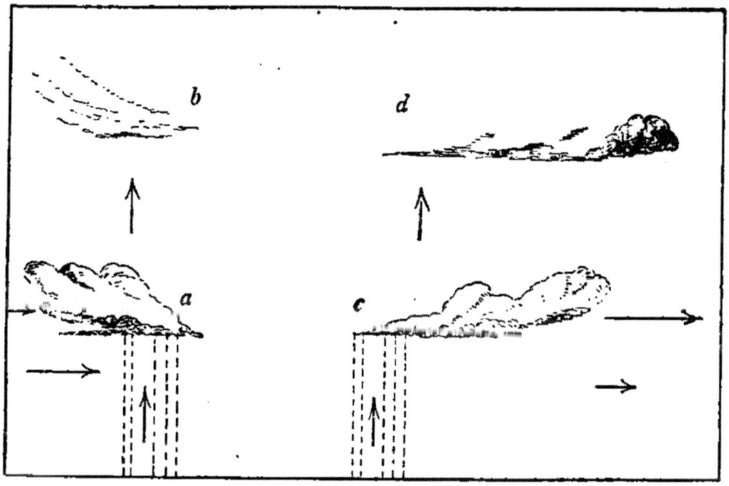

Fig. 11.—Cumulus and cirrus. *a*. Cumulus, surface rapid. *b*. Cirrus, surface rapid. *c*. Cumulus, upper rapid. *d*. Cirrus, upper rapid.

condenses into cloud, like the condensed steam from an engine.

The flat base marks the level where condensation temperature is reached, and the upper rocky summit represents the heads of the air-columns protruding into a cold space. A cumulus is, in fact, the visible capital of an ascensional column of air.

There is one very remarkable feature of all cumulus: it is never seen as such overhead, but only on the horizon, or at a moderate height above it.

The reason is obvious, that as the flat-based mass drifts overhead, the flat under-surface hides the characteristic rocky top, so that we no longer see the typical features of this kind of cloud.

In Northern Europe, and the interior of many continents, cumulus usually only forms during the summer months; for the absolute amount of vapour in the air during the winter months is rarely sufficient to develop a lump of cloud.

Relation to Cirrus.

If the ascensional column is stationary, we get a very curious appearance; the top of the cloud seems to be, and is, in a state of commotion, but still the cloud as a whole does not move in any direction. This is very puzzling at first, and is not uncommon before thunderstorms; but we can readily understand its origin by watching the stationary cloud on a hilltop. Then we see the same contradictory appearance—a cloud in rapid motion, but never moving forwards. The reason is that, as each fresh portion of cloud is projected upwards and blown away by the wind, it is immediately evaporated, but a new column of vapour instantly takes its place. But suppose that, whether stationary or moving, the rising column, after depositing a certain amount of its vapour at one level, continues to rise, it will at length reach a second level, at which the condensation-point of the diminished amount of vapour which it now contains will be again reached, and a second layer of cloud will be formed at a higher level. Sometimes this will be deposited as another cumulus, but more frequently the rising column has so much diminished both in column and upward velocity that the vapour is condensed in a thin, hairy, or curly form, as in Fig. 11, *b*. This is pure cirrus, or curl-cloud, and may often be seen floating above a cumulus. If the upper and lower strata of air are moving at an equal speed, the cumulus, when once formed, sails on without any change of shape from the action of the wind, however much it may alter from any difference in the supply of uprising vapour. But if, as in Fig. 11, *a*, we suppose an ascensional column of air to start from near the earth's surface, and, when it has risen nearly to its condensation-level, to meet an upper current in the same direction as itself, but moving more slowly, then we would get a flat-based cumulus, headed back, as it were, like the cloud, marked *a* in the figure, while the whole mass would move end on with the wind.

In Europe this is practically only found in the rear of cyclones, and we may therefore deduce, from the shape of this cloud, that there the surface is quicker than the upper current.

Suppose, now, that the column of air was attenuated into a thread, then under the same conditions we should get a feathery cirrus marked *b* in the figure, also moving end on. If, now, under similar conditions, the upper current is moving faster than the lower one, we should get in the lower strata a cumulus heading forwards as marked *c*; and if at a higher level the ascensional column was attenuated into a thread, then we should have a light, hairy cirrus marked *d*, moving long-ways with the wind. In this case, sketched from nature, we may note that the foremost curl of the cloud is lumpy like a small cumulus, instead of hairy like the other tufts.

It is from simple illustrations of this sort that we seem to find the connecting link between some forms of cirrus and cumulus. If cumulus is the visible capital of an ascensional column of air, cirrus is the visible form of the condensation of a column attenuated or wire-drawn into a thread. Of the three threads which have formed the cloud *d*, two have died out as hairs, but the third, a little more intense, had had enough volume or energy to form a tiny nubecule.

A common case in which we can trace the gradual development of fibrous cloud into cumulus occurs on any bright morning with a blue sky and heavy dew. When the sun gains a little power, the first threads of cloud formed by the moisture rising into the air are usually condensed into cirrus; these quickly get big and swell, till in a very short time the sky is nearly covered with true cumulus.

The reason why the stripes *a* and *b* are slanting is that in them we have supposed, as is often the case, that the cloud continues to rise after it has begun to condense; in *c* and *d*, on the contrary, the cloud has ceased to rise, and the stripe lies straight.

Ley has shown that descending threadlets of icy particles can be drawn into wispy cirrus, if they fall into a layer of air which moves more or less quickly than themselves. The conditions for such threadlets would be found if the upper outflow from a cyclonic vortex injected vapour into the cold still air of an adjacent anticyclone.

Now for the weather-portent of pure cumulus. In our chapter on prognostics we have shown that cumulus is the specially characteristic cloud of the rear of a cyclone, and it is then often associated with showers. When this occurs we have cumulo-nimbus; that is, rain falling from cumulus to distinguish it from strato-nimbus, where rain falls from stratus cloud. At the edge of anticyclones we often have a fine-weather light cumulus, which is apparently formed by the simple rising of the vapour evaporated from the ground by the heat of the sun, in contradistinction to the cyclone-cumulus, which we may suppose to be produced by the ascensional currents which are generated by the whirling motion of an atmospheric eddy. The form alone does not enable us to say whether a cumulus indicates good or bad weather. This cloud, like every other, must be judged by its antecedents and surroundings.

Cumulus is the almost universal cloud of the tropics, and may indicate either fine weather or non-isobaric rain, according to circumstances.

Festooned Cumulus.

There are two or three modifications of cumulus which it is important to notice, as they throw much light on the nature of cloud-formation. In looking at them, as at all others, we must consider the life-history of a cloud—from what it originates, and to what it develops. Sometimes the lower base of a cumulus assumes a festooned appearance, as in Fig. 12, *b*. In Orkney, this is known as the "pocky cloud," and is there usually followed by a severe gale of wind. In Lancashire, the festoons are called "rain-balls," and are only considered a sign of rain. We have frequently seen them in all parts of the Tropics; but all festooned forms of cloud are unknown in Scandinavia. For a technical international name, Poey has suggested "globo-cumulus," while Ley has proposed the term

FIG. 12.—Festooned cumulus.

"mammato-cumulus," and this latter seems to be very suitable.

The origin of this form will be readily understood by the following example. One summer evening in London, towards sunset, the author saw a flat-based cumulus, like that marked *a* 1 in Fig. 12, suddenly become festooned at the base, and diminished at the top, as marked *a* 2 in the diagram. A few minutes later the whole cloud evaporated, and the succeeding night was fine.

The explanation which immediately suggested itself was, that the ascensional current which formed the flat-based cumulus had suddenly failed, and that the festoons were simply the masses of vapour falling downwards for want of support.

Another very striking case is marked *b* in the figure,

and was observed before a shower. Here a detached cumulus was observed to form, first, festoons, and then they in turn degraded into raggy cloud; the whole disappeared very shortly, but were quickly followed by fresh rain-bearing clouds. The impression was that the festoons were formed by a sudden drop of the cloud, and that the rag was produced when the drop was less sudden. The appearance is, unfortunately, not well rendered in the diagram. From many similar instances, we are led to the conclusion that the constant condition necessary to form festoons is the sudden failure of an upward current of air, and then we can readily see why they should prognosticate a storm in some cases, and only rain in others.

Before many squalls or showers, we are all familiar with the short abortive gusts which so often precede them. Now, we have only to assume that the ascensional uptake in front of the main body of the shower is as unsteady as the surface-wind, and we have at once all the conditions necessary to form festoons. All observers are agreed that they are usually formed at the edges of cloud-masses. In the case of rain or thunder, they ordinarily appear just before or after the rain; but when a gale follows some time afterwards, the festoon must have been formed by some local squall or shower, that bore some relation to the disturbed weather which produced the gale. In Orkney the festoons are usually seen with the squalls of a north-west wind in rear of a cyclone; the storm they prognosticate belongs to another cyclone, which there usually follows quickly behind the first. In the tropics, of course, festoons are always associated with non-isobaric rain or thunderstorms.

Degraded Cumulus.

A very similar line of argument applies to another well-known sign of rain—the appearance of cloud shown in Fig. 13, *a*, where a thin stripe of cloud seems to cross a well-formed cumulus.

This is a foreshortened view of a cumulus (*b*), and a degraded patch of cloud (*c*). Sometimes, but most unfortunately, this cloud is called cumulo-stratus, as by Howard and others. We shall presently see, however,

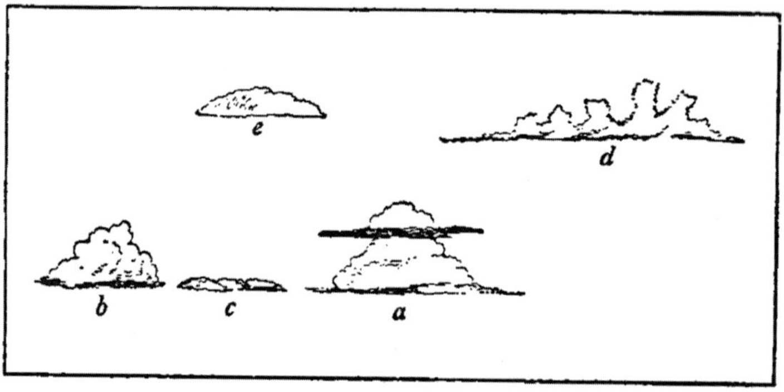

Fig. 13.—Cumulus, degraded cumulus, and line cumulus. *a.* Cumulus crossed by another cloud. *b, c.* The same from another side. *d.* Line cumulus, or high cumulus. *e.* Degraded cumulus, lens-shaped.

that it has nothing in common with true stratus, but is a mixture of pure cumulus with a degraded patch of cloud. The origin of this cloud is very simple. We have shown in the preceding diagram that a detached cumulus (Fig. 12, *a* 1) can become degraded into a flat, thin mass, with festooned base; but in certain conditions the failure of the rising current takes place more gradually, and then the base of the mass remains flat instead of becoming festooned.

This, too, is a sign of rain for the same reason as in the former instance. The existence of failing or abortive rising currents is of itself a sign of disturbed weather, and is really more of an accompaniment than a prognostic of rain. This cloud is common in the equatorial doldrums, and in any other part of the world where showers fall from cumulus. Thus we see that festooned, raggy, and this streaked cumulus are all associates of rain, and for similar reasons. A very similar degraded cumulus-patch is very common during the finest weather in the trade-wind districts. The small isolated patches of cumulus, which are so common there, often seem to lose so much of their rising impulse that a rocky top cannot form, but at the same time the stoppage is not so sudden, or the cloud so heavy, as to develop festoons. Then we get a cloud nearly flat below, with a smooth round surface above, like a plano-convex lens, as in Fig. 13, *e*. But, as Ley finds an almost identical form as the embryo of a cumulus whose rising force is very weak, we must judge the import of this, as of every other cloud, by its surroundings.

Minor Varieties.

Another form of cumulus is developed almost at the level of cirrus, in long thin lines made up of little heads of condensed vapour, sometimes called thunder-heads. This is shown at the right-hand top corner of Fig. 13, *d*. It is only noticed here to guard against its being called cirro-cumulus. In practice this cloud is almost invariably produced in front of thunderstorms, and it is difficult to see how it can be formed otherwise than by assuming the air to rise in a thin vertical curtain. In our chapter on line-thunderstorms, we shall find from other reasons that the air really does sometimes rise in long narrow sheets. This is the cumulus simplex of Weilbach and rain-cumulus of Howard; it has also been called high cumulus, line-cumulus, and turreted cumulus.

There is one other variety of cumulus, which need only be mentioned here.

Sometimes the top of the cumulus becomes hairy, as if it had been combed out; this cannot be explained, but is usually seen over heavy rain. But occasionally this peculiar process on the top of a rainy cumulus develops a sort of flat sheet of cloud, apparently touching the summit, and the cloud may conveniently be called cumulo-stratus.

Stratus.

We now come to the second variety of clouds, to which the name of stratus is applied, because it always lies in a thin horizontal layer, like a stratum of rock or clay. Pure stratus has no sign of any hairy or thread-like structure except at the edges, for a stratum which shows much marking would be cirro-stratus, and has quite a different origin. Pure stratus is essentially a fine-weather cloud, and is especially characteristic of anticyclones. One very beautiful variety is often seen during a fine night, when the cloud forms thin broken flakes, something like mackerel sky, from which, however it is really quite distinct.

In Howard's original work on clouds, "stratus" was

applied to ground-mist, but that idea is now entirely discarded by all meteorologists. What we call pure stratus is the "strato-pallium" of Weilbach, and the "stratus" of Hildebrandson. The origin of this cloud seems to be that when the air is tolerably still, and radiation is going on, the general mass of the air gets gradually cooler, till at last the temperature is reached at which some stratum touches the dew-point, and therefore condenses its moisture into cloud. Sometimes the cloud is formed by rising fog.

This at once explains both why the stratum of clouds should be flat and thin, and why this form of cloud should be characteristic of anticyclones. We can also understand why, under these conditions, the sky sometimes becomes overcast almost instantaneously. Very often a mass of fine-weather stratus is uniform in the centre, but hairy or striated at the edges, and we get a cloud indistinguishable by form alone from some kinds of strato-cirrus, though very different in origin and surroundings. Sometimes the lower surface of a sheet of cloud is festooned for a short time like the flat base of a cumulus. The cloud is probably not then pure radiation stratus, but a smooth form of strato-cumulus, which, by sudden failure of the generating current, begins to fall in lumps just like the festooned cumulus before described.

Cirrus.

The third primary form of cloud is cirrus, a word taken from the Latin, and meaning literally "a curl of hair." We have already explained the origin of pure cirrus and its relation to pure cumulus, together with the rudimentary idea of the formation of a stripe of cloud from a current of vapour-laden air, which rises in currents of different velocities, but in the same direction.

Cirrus-Stripes.

When cirrus rises irregularly, and appears not to be all at the same level, we have seen that it is then pure cirrus; but there is a modified form, in which more or less of the sky is covered with long thin stripes of cirrus, all apparently at the same level. Technically this is known as cirro-filum (literally "hair-thread"), a name first suggested by Mr. Ley, and the term is suitable for international use; but we shall call them cirrus-stripes. As these are by far the most important form of cirrus for forecasting purposes, we shall devote several paragraphs to their consideration.

First, as to their origin. We have already explained how a stripe can be formed which moves end on to the wind that is propelling it, but most frequently we see the curious spectacle of a long stripe of cloud moving either broadside on or obliquely to its length. As we must suppose that a stripe always sails with the wind in which it floats, we have to find out how a stripe can be formed which moves across its length. At first sight this is one of the most puzzling phases of cloud-motion. These formations of cloud are, however, exactly analogous to the smoke left by a steamer running before the wind. If she runs faster than the wind, her smoke trails behind; but if the wind blows faster than she steams, then the smoke is blown forwards in front of her. But now, suppose her to be heading to the east with a south-west wind; it is obvious, from Fig. 14, that her smoke would lie in a stripe bearing somewhere between north-west and south-east, and would drift towards the north-east, that is nearly at right angles to its length. The smoke that left the funnel when the steamer was at A would have been blown to C by the time she had reached B, while that at B would

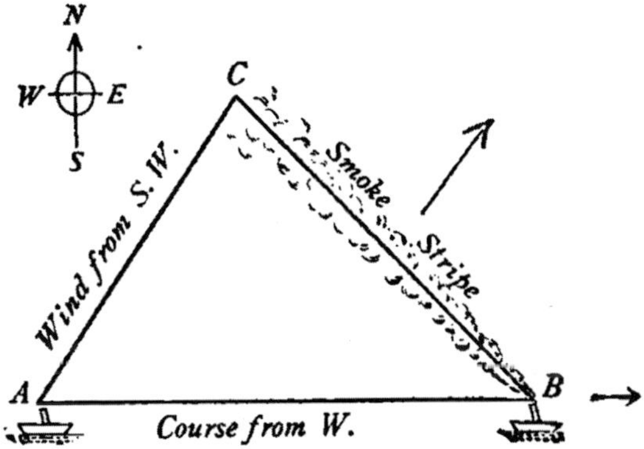

FIG. 14.—Formation of cloud-stripes.

be just leaving the smoke-stack; so that the whole line of smoke would lie from B to C, but drift from south-west with the wind. The angle the smoke forms with the course of the ship obviously depends on the speed of the ship and the velocity of the wind, so much so that we have used measurements of the angle A B C to determine the velocity of the wind at sea.

Now, this is exactly what happens in nature. The ascensional column of moist air, which will eventually form a cumulus, starts from near the earth's surface, drifting with the wind which blows there; when it arrives at a certain height, it meets an upper current moving in a different direction to that on the surface, and probably begins to condense there. The stripe which would be formed under these circumstances would behave exactly like the smoke of a steamer; that is to say, it would lie obliquely to the wind which was driving it.

The direction of a stripe is sometimes called the direction of its filature, but we shall employ the less technical term of "the lie of the stripe." The triangle A B C, Fig. 14, is called the triangle of filature, and it is evident, from the nature of the composition of velocities, that the precise direction of the lie of the stripe depends on the relation of the velocities of the upper and lower currents.

If the stripes were caused by descending threadlets of ice or snow, the above principles of stripe-formation would equally hold good.

Lie and Motion of Stripes.

Before explaining the nature of the upper currents in cyclones and anticyclones, we must first explain how to find out both the lie of a stripe and the direction in which it is moving, as both these points are important, and both rather difficult to observe. If the sky is well covered with stripes, we find that when we look at them lengthways they appear to converge towards a point on the horizon, while, if viewed transversely or in profile, they appear arched. The convergence, of course, is a matter of perspective.

One great peculiarity of stripes is that, while in their simplest form the threads of which they are composed lie in the direction of its length, as marked *a*, *e*, in Fig. 15, sometimes the whole stripe is made up of a series of cross-bars, and the stripe is then said to be "striated" (Fig. 15, *b* and *d*). Most frequently these bars, or striæ, are at right angles to the length of the stripe, but they are also sometimes oblique to the lie of the stripe.

Whenever the cloud is observed, the points to be noted are—(1) the direction or lie of the stripe; (2) the direction of the striæ; and (3) the direction in which the

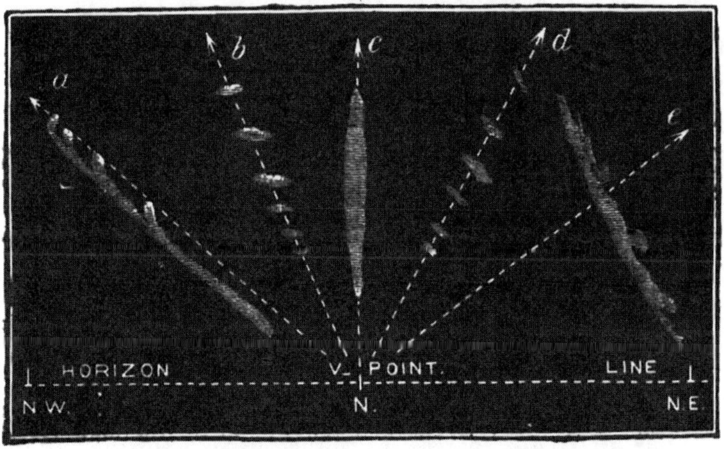

FIG. 15.—Diagram illustrating cloud-perspective.

stripe as a whole is moving. Of these the first and third are the most important, the direction of the striæ being only secondary.

Cases, however, occur in which the whole sky is covered with a cloud reticulated like a chess-board, and it is then difficult to say which is the primary filature of the cloud. We will first consider how to determine the direction of the stripes and striæ, as that is far easier than to discover their motion. The method of doing so is based on the fundamental principle of perspective—that a line drawn from the observer to the point on the horizon towards which parallel lines converge, gives the lie of the parallel lines. Thus suppose, as in Fig. 15, that an observer looking north saw stripes *a*, *b*, *c*, and *d* converging on the north point of the horizon, he would conclude that what he saw was the perspective view of four stripes lying in a direction given by a line drawn from himself northwards—that is, north and south. On the one hand, he would know that the striæ of stripe *b* were lying east and west—that is, at right angles with the filature; for these striæ converge nowhere, but are parallel to the horizon-line when looking north. On the other hand, he would see that the striæ of stripe *d* converge on the north-east point of the horizon, and that, therefore, the striæ lie north-east and south-west, while the stripe as a whole points north and south. For a similar reason, he would know that the single stripe *e* lay also from north-east to south-west.

The above are very striking instances of the deceptive nature of perspective, for in stripe *b*, where the striæ are really at right angles to the stripe, they appear oblique; while in stripe *d*, where they are really oblique, they look as if they were nearly at right angles to the lie of the cirrus. The point on the horizon towards which stripes or striæ converge, is called their vanishing-point, or, more shortly, their V-point. Some observers, however, prefer the term "radiation-point," and talk of the radiation of cirrus. Had the stripes been viewed looking straight either east or west, they would have presented the appearance of an arch, whose flat top bore due east or west, and whose ends pointed north and south.

A very simple way of learning cloud-perspective is to stand at the end of a long room and assume that you are looking north; then you see at once that the lines of the two cornices, on either side of the room, converge towards the north; and, if you can suppose striated lines like those of Fig. 15 to be painted on the ceiling lengthways to the room, you will see that the striæ would converge in the manner shown in that diagram.

Now for the more difficult question of finding the motion of the stripe.

Usually, the motion of the stripe is not in the direction of its length. Frequently it moves broadside on—that is, at right angles to its length, but more frequently at an oblique angle to its length.

The most accurate observations can be made when clouds are exactly overhead. Then, of course, there is no delusive perspective, and the direction from which the cloud comes is the direction wanted. A stripe moving obliquely to its length is, however, always a difficult subject.

In most cases, however, it is impossible to catch a cloud just overhead, and even then it is most inconvenient to observe; so that as a rule we must use clouds at a moderate elevation, and allow for perspective. The perspective of motion is, of course, the same as the perspective of shape; that is, the point from which the motion appears to diverge is the point from which the cloud is really moving, and the point to which the motion converges is the point to which the clouds are travelling.

For instance, take an easy case, when fine detached cumulus is moving rapidly from the north. If we look anywhere towards the north-west, a glance at Fig. 15 will show that a cloud moving along one of the lines diverging from the V-point would appear to have some motion from the east as well as north; in fact, would look like a north-east motion. Conversely, when looking towards the north-east at a cloud like *d*, the motion would appear to be somewhere from the north-west; but if looking due north, as at *c*, the cloud would appear to rise straight out of the horizon. The rule therefore is, watch the point from which the motion of the clouds seems to diverge, and that is the direction from which they are really moving. Sometimes it is more convenient to determine the point towards which the clouds converge, the V-point giving then the direction towards which the motion is. If possible, however, the point of divergence should be selected as being the easier to observe.

The reason why the motion of cirrus-stripes is so much more difficult to determine than the direction of their filature is that, being very narrow, you get only a very short line from which to estimate the V-point, if the motion is not in the direction of the stripe's length. For instance, in Fig. 15, where the dotted lines denote the divergence motion of the different stripes, suppose that we had been able to watch the motion of the stripe *a* as it drifted past a star or any fixed point, we should find that the line of motion was coincident with the length. The motion of the stripe is, therefore, from the V-point—that is, from the north. On the other hand, if we watched the stripe *e*, which is really moving only partially sideways, though apparently it moves at right angles to its length, we should only have the short length of the dotted line which passes through the stripe to produce by eye till it cuts the horizon from which to estimate the V-point.

Then consider the increased complication when the stripe is striated. In stripe *b*, though the motion of the stripe coincides with its length, each bar, or stria, would appear to lie obliquely to the line of its motion, though really moving at right angles to its length. On the other hand, the striæ of stripe *d* appear to move at right angles to their length, when they really move obliquely.

Another and sometimes even greater difficulty arises from the changes which are going on in the cloud itself. If we take two photographs of a cloud at an interval of only three minutes, it is sometimes impossible to identify the same portion of the cloud in the two pictures. When, therefore, we get the still more complicated case of an obliquely striated stripe, which moves very slowly at a considerable angle to its filature, we can readily understand that the true character of its motion can only be determined under favourable circumstances and by a skilful observer.

Such are the fundamental principles on which all observations on cloud-motion depend. The observer must not be deterred by difficulties at first starting. If he begin by taking simple cases of fast-moving clouds, and then, after he has fully realized the meaning and importance of the V-points, tries more difficult cases, he will soon attain such proficiency as will enable him to make valuable observations in the most recent branches of modern cloud-science.

If possible, the velocity of the upper clouds should be noted, for quickly moving upper clouds are a sign of much worse weather than slow-going ones.

Relation to Cyclones and Anticyclones.

It has been found, as a matter of observation, by Ley and others, that the lie of cirrus-stripes bears a tolerably constant relation to the shape of isobars over the locality where they are seen. It is from this circumstance that cirrus derives its great forecasting value; also from the fact that it is the first cloud which appears in a sky which had been previously blue. Hildebrandson finds the following deviation of the stripe to the isobar out of 171 observations. The stripes whose angles of deviation are greater than 45° are, of course, less nearly parallel to the isobar than those whose angle is less than 45°.

Deviation.	Anticyclones (maxima).	Cyclones (minima). Front.	Cyclones (minima). Rear.	Wedge isobars.	Totals.
Angle greater than 45° ...	58	5	8	3	74
Angle less than 45° ...	12	18	29	38	97
Total	70	23	37	41	171

Consequently, cirrus-stripes lie in regions of maximum pressure most often nearly perpendicular to the isobar, while round minima and along wedges they are more nearly parallel to the isobars.

To explain the reason of this, we must now show the relation of the upper to the lower currents in cyclones and anticyclones. Our knowledge of the upper currents has been deduced entirely from cirrus observations.

VERTICAL SUCCESSION OF AIR-CURRENTS.

In Fig. 16 we give a diagram of the surface and highest currents in both a cyclone and an anticyclone, as deduced by Ley, Loomis, and Hildebrandson of Upsala.

The solid arrows denote the surface-winds, while the highest currents are given by the dotted arrows; and the two arrows are supposed to diverge from the point of observation. In a few places, where the velocity of the two currents is usually different, we have drawn the respective arrow of different lengths, otherwise the arrows

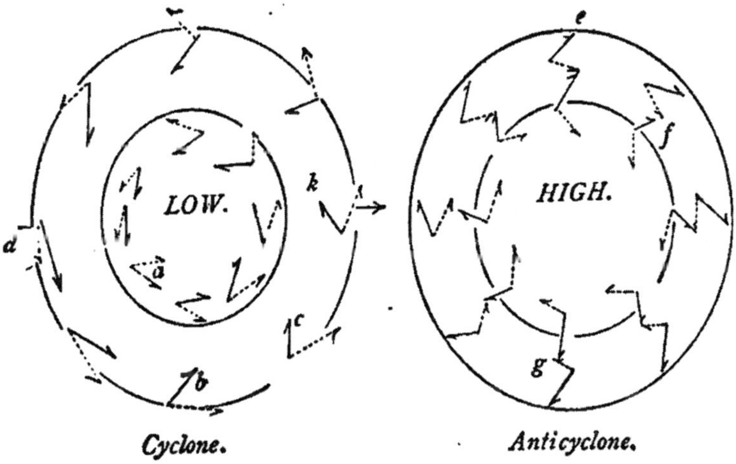

FIG. 16.—Surface and highest currents over cyclones and anticyclones.

must be supposed to give only the relative directions of the winds.

First for the cyclone. Let us consider the nature of the surface-winds, as shown by the solid arrows. We see at once that on the whole the direction of the surface-wind may be described as an ingoing spiral, more incurved in the right front than in any other portion, and that in all parts the wind is a little less incurved the nearer we approach the centre. In the diagram we have assumed that the cyclone is moving due west, and that it is also truly circular. The dotted arrows, on the contrary, show that the wind in the upper strata blows in an irregular spiral outwards; and that, while in front of the cyclone the upper winds are very much inclined outwards, in rear they are very nearly parallel to the surface-currents.

If we had put in arrows to show the direction of the lower cloud which floats from 6000 to 8000 feet above the earth, we should have found that intermediate layer moving almost exactly parallel to the isobars—that is, nearly in a circle. We also know, from other observations, that the upper cirrus-current in front of a cyclone is much nearer the surface than the same current in rear of the centre. Now, when we come to look at the direction of cirrus-stripes, as given approximately by an imaginary line drawn through the points of each pair of arrows, we see at once that in the outer circle especially, as at *b*, *c*, *d*, the cirrus-stripes will lie at less than 45° to the isobars, at the point from which the arrows diverge in the diagram if stripes are formed as shown in Fig. 14.

In practise the stripes are more nearly parallel to the isobars than would appear from the generalized diagrams, as the majority of cyclones are not circular, but oval; and that the effect of an intermediate current of air less incurved towards the centre, would be to make the stripe less at right angles to the isobar than would appear from the diagram. Note, also, that in front the surface-winds are slower than the upper ones, while in rear the surface are the quicker; also especially that in every portion with one exception, as at *a*, the upper current is always more veered than the lower one—that is to say, that if the surface is east, the upper will be more south of east, and if the surface is west, the upper will be more north of west. Or we may put it thus: stand with your back to the wind, and the upper currents always come more from the left, and the higher the currents the greater the amount of veering. This is the almost universal law of upper winds in the northern hemisphere for clouds at all levels.

If the surface is east, and the low clouds south, the higher cirrus will be from some point west of south; but if, with the same surface east wind, we saw cirrus driving from south, we should know that the intermediate currents of wind cannot be veered so much, but must come from some point of south-east. This sequence, which we will call the law of vertical succession of upper currents, is another of the fundamental principles of meteorology.

In certain well-defined cases we find an apparently anomalous sequence, but these need not be described in an elementary work.

Then for the anticyclone. Referring to Fig. 16 again, we see that, as a general rule, the surface and highest winds are much more opposed to one another than in cyclones, and therefore in most cases the cirrus-stripes will be more nearly perpendicular to the isobars, as at *e*, *f*, *g*. Taking a general view of the surface-winds, we may say that on the whole they blow spirally outwards, in the direction of the motion of the hands of a watch; that they are less square to the isobars the further they are from the centre, but that they are always more nearly perpendicular to the isobar than the wind in any portion of a cyclone.

The upper currents, on the contrary, blow spirally inwards, also in the direction of the watch-hands, and also much inclined to the isobars.

Now, taking a general view of the relation of stripes to isobars, we must not expect the lie of the stripe to be more than a moderately good guide to the lie of the isobars. Independently of the fact that there is no hard line of demarcation between a cyclone and an anticyclone, across which the stripes should suddenly alter from parallel to transverse to the isobars, it is manifest that when so much depends on the relative velocities of the upper and lower currents, much variation in the lie of the stripes must be expected.

But besides this, the author has sometimes found that the intermediate current between the surface and the cirrus materially affects the lie of the clouds, and that the lie of some stripes cannot be explained on this principle at all; but, in spite of all this, the above generalizations on the lie of stripes are very valuable.

The following example, as observed by M. Phillipe Weilbach, Copenhagen, illustrates our general principles. Fig. 17 represents a portion of the cirrus-stripes as seen at 1 p.m. on October 27, 1880, at Copenhagen, foretelling the bad weather which occurred on the following days. The sky generally would be said to be covered with striated cirrus-stripes, really forming a thin layer of cirro-stratus, which appeared in great quantity and covered the whole of the eastern sky from the zenith to the horizon, forming long bands, which diverged from the north-north-west tangentially to the isobars.

The barometer was at 29·6 ins. (752 mm.); the wind blew lightly from the north, while the relative humidity was only sixty-seven per cent.; the rest of the sky was sparsely covered with fibrous clouds of divers forms. For several hours the bands of cirrus, striated by fine lines, moved

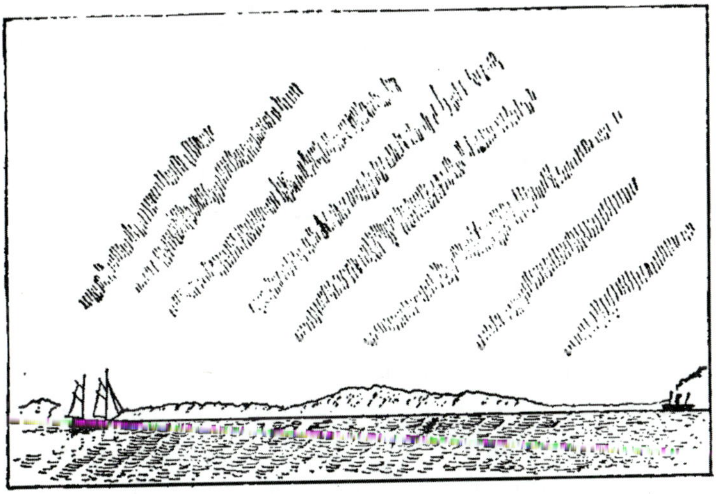

FIG. 17.—Converging striated cirrus-stripes.

from the north-west; then the sky at Copenhagen became clear for some time.

In the afternoon the preceding appearance was replaced by a *ciel pommelé* (dappled sky), clearly marked, but of a rather heavy aspect, which also moved slowly from the north-west.

At the same time the telegraph gave notice of a storm from the south-east over Ireland, and the following day the tempest raged with heavy snow over Denmark, while the depression moved to the south of that country. The landscape in the figure is looking up the Sound.

All this is very easily explained. When the cirrus was first observed, Denmark was under the influence of the rear of a cyclone, rather than of the wedge, which lay a little farther to the west, while a new cyclone was forming behind the wedge. The isobars would, of course, lie from north-west to south-east, nearly the same as the cirrus. The episode of the sky becoming perfectly clear after the first indications of dangerous cirrus is very common, and seems due to the first cat's-paws, as it were, of the oncoming cyclone developing cirrus, and then failing, so that the sky clears again.

Fine Weather and Dangerous Cirrus.

There are many forms of pure, hairy cirrus that indicate fine weather all over the world; while others, such as "mare's tails," "cat's tails," "goat's hair," "sea-grass," and "gashes" (*balafres*), etc., are forerunners of bad weather in every country.

In England "mare's tails" usually portend wind, and "goat's hair" only rain; while "mare's tails," "cat's tails," and *balafres* precede every hurricane in the tropics. "Mare's tails" are long, straight fibres of grey cirrus; "cat's tails" are a denser bundle, often slightly striated, so as to look like brindlings on the tail; "goat's hair" is a short bundle of white cirrus-hairs; while "sea-grass" and *balafres* are both somewhat similar to the above allied forms.

These all represent forms of cirrus at the outskirts of

cyclones, intermediate between the pure wisp of fine weather and true cirrus-stripes, with the exception of "goat's hair," which is a form of cirrification on the top of rain-bearing cumulus; but in every country there are sometimes illusory forms, which it is very difficult at first to connect with good or bad weather. In England, on a fine summer day, detached cumulus, which has formed during the afternoon, will become very small and disappear towards sunset; and straight fibres of cirrus will gradually appear in the sky, which by form alone are indistinguishable from "mare's tails" and similar forms of cirrus which presage wind. All over the tropics the typical sky by day is lumps of cumulus floating below wisps of cirrus, and, without considering their surroundings, the latter might be thought to be indicative of coming danger. Clouds must always be judged by their antecedents and surroundings. The gradual growth of cirrus-fibres after cumulus in England, and the general appearance of the weather and the diurnal fall of the wind, will usually prevent any mistake from being made between the "mare's tails" of a summer evening and the similar cloud which streaks the sky on a windy, gusty day.

The "cat's tails" which precede a tropical hurricane do not disappear shortly after sunset like the ordinary wisps, and when combined with a slow, steady fall of the barometer, a dangerous storm is certainly indicated.

This is a simple case of what we find throughout all cloud-lore—that the same cloud does not always indicate the same weather, even in the same country. Here the explanation is doubtful. Some think that the essential to form a fibre of cirrus is a thin thread of damp air rising slowly into a current different in speed or direction from that in which the air started.

If the rising impulse is merely the effect of the sun heating air near the ground, the resulting wisp is a fine-weather cirrus; but if, on the contrary, the ascent of air is due to the upward impulse of a cyclone, then the bundle of cirrus-fibres indicates wind and rain.

Others consider that the advent of damp upper currents in front of a cyclone induce the condensation of vapour at a high level into icy particles, which latter are drawn into wisps of cirrus as they descend into lower strata.

We believe that cirrus may be formed by both methods, but it is impossible to pronounce definitely on the subject, till we know more of the mechanism of a cyclone.

Cirro-Stratus.

We now come to the composite forms of clouds, and here, unfortunately, we find the utmost confusion in the words applied by different meteorologists to the same clouds. We will first begin with cirro-stratus. By this we mean a thin stratum of cloud which, instead of being uniform like pure stratus, is composed of fibres of cirrus in any complexity, but not of streaked, fretted, or speckled nubecules.

Sometimes the fibres of cirrus interlace, and give this cloud a reticulated appearance like a woven cloth, and the variety of forms is unlimited. The cloud we call cirro-stratus is practically identical with what Howard and Hildebrandson call "cirro-stratus," and almost co-extensive with the *cirro-velum* of Ley. As to the origin of cirro-stratus, we can say little with certainty. As a matter of observation, it is usually formed in front of cyclones or secondaries. When the sun or moon shine through it, we generally find that a halo is formed, and then we may conclude with certainty that it is composed of frozen particles of vapour. The difficulty is to explain the innumerable forms which it assumes, and the rapid changes which it undergoes. Though we are obliged to employ the word *strato* to describe this cloud, because it forms a thin layer, it is extremely doubtful whether its formation has much in common with that of pure stratus, which we have seen is due to the radiation of anticyclones. It does, however, seem to have something in common with pure cirrus, and still more with cirrus-stripes; but we cannot say why the front of a cyclone should develop stratiform, and the rear cumulo form, clouds.

Sometimes cirro-stratus is formed lower down, and more compact in structure, when it should be called strato-cirrus. This is unknown in Scandinavia, but quite common in some parts of the tropics.

Origin of Striæ.

With regard to the striæ which we find both in cirrus-stripes and in cirro-stratus, the only reasonable suggestion which has been proposed to account for their formation is, that a stripe, or a thin stratum of ice-dust, may sometimes be supposed to be relatively at rest to a wind more rapid than itself, which may strike it suddenly. Then we can conceive that a smooth layer of cloud might be furrowed into small waves at right angles to the wind; but this, of course, would only account for striæ square to the stripe, and not for oblique markings. We often see an apparently structureless patch of cirro-stratus suddenly become striated, as if a cat's paw of wind had blown on it like a gust on a pond.

But, as a matter of fact, striæ are as often as not oblique to the lie of a stripe, and to the direction of the motion of cirro-stratus. Here also the only rational suggestion is that the oblique striations are in some way the effect of an upper current, which moves in a different direction to that on the surface, and forms cloud rolls.

There is certainly something to the eye about the sideways motion of some cirrus-stripes that is not the same as the drive of a detached cumulus before the wind. If one is really propagated by a dynamical disturbance, while the other merely floats in an air-current, the difference would probably be explained. If this can ever be satisfactorily worked out, we should get the motion of higher currents more accurately than at present, for now we always assume the motion of a cloud is the same as that of the wind which drives it along.

Sometimes a succession of rising threads of air, one behind the other, form nearly vertical parallel fibres of cirrus, which must not be mistaken for horizontal striæ. All observers are agreed that the fact of striation, or reticulation, is of no practical importance in forecasting weather from clouds, so that we do not make definite varieties of these forms.

Cirro-Cumulus.

The next great class of compounds is cirro-cumulus. By this we mean a broken layer of cloud, at a high or middle level, of which the component masses are not fibrous like cirro-stratus, but more or less rounded or rolled, though without any of the rocky look of pure cumulus.

For this reason the term cirro-cumulus is to a certain extent unfortunate; but we are almost obliged to use the word, so as not to introduce new expressions, and, so long as it is conventionally recognized what kind of cloud is meant by cirro-cumulus, it does not so much matter if the word is not quite logical. The misfortune of the word "cirro-cumulus" is that, even excluding the small high cumulus that sometimes grows out of hairy cirrus, and which we have described as linear, or high, cumulus, there are still two rather distinct forms, to either of which the definition we have given of cirro-cumulus applies.

The first kind, and far the commoner all over the world, is composed of rolled masses of cloud, with a fleecy appearance, that are universally known in different languages as "wool-pack," "sheep," "lambs," or by similar terms. This is the cirro-cumulus of Fitzroy, Weilbach, Hildebrandson, and of Howard. The clouds called *nubes hiemales* by Weilbach, are a variety of this type that is formed with great persistency over Scandinavia and Northern Europe during the cold season. The thin layer of cloud is then at a moderate altitude, and tends to arrange itself in long parallel bands of quickly moving, fleecy masses.

It is extremely difficult to render that kind of cloud in an engraving. Fig. 18 is, however, a moderately successful attempt to reproduce a photograph of a fleecy sky. There, as always, the cloud has a more or less pronounced tendency to arrange itself along two lines—

FIG. 18.—Fleecy cirro-cumulus.

one for the length of the bands; the other for the lie of the striæ. Sometimes the effect of these two crossing lines is to give the individual nubicules which compose the whole a square or lozenge shape, and the whole sky the appearance of a gigantic chess-board.

We can say little with certainty as to the formation of this fleecy sky, though, in a general way, there seems to be little doubt that both the woolly look and the striation are due to the contact and rolling friction of two layers of air moving in different directions. Fleecy clouds, though apparently so different in form, are really not very far removed from wispy cirro-stratus. We often see in England wispy clouds develop rapidly into fleecy ones for a few minutes, and then back again into wisps and curls; but, as a rule, cirro-stratus develops into strato-cumulus, and is practically a sign of worse weather than fleecy cirro-cumulus.

We know by observation that fleecy cirro-cumulus is chiefly formed in the temperate zone on the edges of anticyclones, and also before thunderstorms and some forms of non-isobaric rain. These are both cases in which there would be upper currents varying much in direction from the surface-winds, while the rapidity of motion would depend upon circumstances. This enables us to explain the following set of widely reputed prognostics.

> "If woolly fleeces spread the heavenly way,
> Be sure no rain disturbs the summer's day."

Or the provincial French saying, "El ciel pecoun promête un bel matin." But, on the other hand, Virgi ("Georg." i. 397) considers it a sign of rain if it should happen that—

> "Tenuia . . . lanæ per cælum vellera ferri."

And so in the neighbourhood of Pisa they say, "Cielo a pecorelle, Acqua a catinelle;" and in the Tyrol, "Sind Morgens Himmelschäflein, wird's Nachmittags hageln oder schnei'n;" and in France they have a proverb contrary to the one we have first quoted:

> "Temps pommelé, fille fardée,
> Ne sont pas de longue durée."

The term "dappled sky" (*ciel pommelé*) is a little equivocal, and might refer to the other form of cirro-cumulus, known in Northern Europe as "mackerel sky."

Anyhow, we have to reconcile an apparently contradictory set of prognostics. The reason appears to be that in Northern Europe rain is chiefly cyclonic, and therefore rarely preceded by fleecy cirro-cumulus, so that the appearance of that cloud denotes the edge of an anticyclone, and fine weather for a day at least. In Central and Southern Europe, on the contrary, fleecy clouds are usually formed in front of secondaries, thunderstorms, and non-isobaric rains, so that their cirro-cumulus is a sign of approaching rain. We can readily imagine that, both at the edges of anticyclones and in front of secondaries, thunderstorms, etc., we have upper currents moving in very different directions to those on the surface, with a layer of cloud between them, though the origin of the condensed vapour is not the same. In anticyclones the vapour probably rises from evaporation, till it reaches an altitude where the temperature falls to the dew-point; in secondary cyclones, etc., the upward impulse is due to the dynamical properties of cyclonic or other motion.

This is exactly analogous to the difference between the wispy cirrus formed of an evening at the edges of anticyclones in fine weather, and the same cloud which precedes a dangerous storm.

In practice the surroundings are so different that the

apparent similarity of names rarely misleads the most ordinary observer.

The second chief variety of cirro-cumulus is composed of rounded and isolated nubicules without any fleecy texture.

This is the well-known "mackerel sky" of Northern Europe; and when the cloudlets are a little angular, we get a form called "mackerel-scales." We may call this hard cirro-cumulus, to distinguish it from the fleecy form of the same generic name. While fleecy cloud is one of the commonest, mackerel is one of the rarest skies, so that we have not got a sufficient number of observations to correlate these isolated cloudlets with any particular form of isobars or kind of rain.

However, all weather-lore connects mackerel with fine weather, for even in rainy Ireland we find the saying, "Mackerel sky, twelve hours dry." Why this should be the case we are unable to say, but there is no doubt about the fact.

In a still rarer form of cirro-cumulus, the lower surface of the general cloud-stratum exhibits very small pendulous protuberances, resembling sacks or bags, by which a part or even the whole sky is festooned. Ley calls this *cirro-velum mammatum*, but we may call it festooned cirro-cumulus. When, near sunset in the tropics, these festoons take up a rosy tint, and hang like pink grapes in a serene sky, these clouds can scarcely be surpassed for beauty.

Sometimes a more compact form of fleecy cirro-cumulus is found at a lower level, when the cloud may be more appropriately reported as cumulo-cirrus, so as to indicate its lower level. This apparent multiplication of cloud-names is forced on us by the necessity of giving some idea of height in reports of the motion of the upper currents. For instance, on the west edge of an anticyclone low cumulo-cirrus might be moving from south, whilst the higher cirro-cumulus would come from the south-west; so that observations which reported cirro-cumulus and cumulo-cirrus indiscriminately would lead to a discordant or erroneous view of the general circulation of the air in an anticyclone.

STRATO-CUMULUS.

Another of the great series of compounds is strato-cumulus. By this we mean a large mass of cloud, forming a layer, which is not sufficiently uniform to be called stratus, and not sufficiently rocky to be called cumulus. This is the cumulo-stratus of Fitzroy. Howard's cumulo-stratus is not a true variety of cloud at all, but a compound of a thin patch of cirro-stratus, resting either on the top of a cumulus or crossing an isolated lump of cumulus, as in Fig. 13, *a*. The origin of the name is obvious. The general mass of the cloud is a layer, and therefore the name must contain the word *strato*, while the components are lumpy, and it must therefore contain the word *cumulo*.

This form of cloud is typical of a cyclone-front in Great Britain. We can trace its gradual development in all stages. Cirrus-stripes first get thicker and lower, so as to form cirro-stratus. As we get nearer the rainy portion of the cyclone, the cirro-stratus loses its fibrous texture, becomes still denser and nearer the earth's surface, till at last all trace of structure is lost in the irregular, shapeless masses of cloud which cover the whole sky. Still later, the cloud gets even lower and blacker, till rain ultimately begins to fall. Then the cloud would be called nimbus, because it forms a layer and precipitates rain. Sometimes, when the sky breaks for a moment, we get a glimpse at the composition of this cloud; we then see that it differs much from pure rocky cumulus, by reason of its flatness and comparative thinness. We must, in fact, look at strato-cumulus as a development of cirro-stratus, and not as an ally or hybrid of cumulus, though we have to use the word "cumulus" in composition.

The point which we cannot altogether explain is, why in front of the cyclone's trough the clouds should have such a marked tendency to form stratus, while in rear the rising currents take the form of well-defined columns, and produce rocky cumulus. This points to some difference of symmetry between these two portions of a cyclone, and the only suggestion which we can make is, that perhaps it may be partly due to the upper currents in front of the trough being much more opposed to those on the surface than those in rear of the centre, which are nearly parallel to the lower winds; and partly to the forward motion of the cyclone, as a whole, meeting the incurving winds in front, and running away from them in rear of the disturbance.

Another form of strato-cumulus is very common in the tropics. The component masses of cloud are more isolated than in Great Britain, and so thin that when seen in perspective each only looks like a dark thin bar, and, with the brighter intervening spaces, the whole sky near the horizon is striped like a Venetian blind.

Nearer overhead we see only the irregular flat base of scattered clouds, without any trace of arrangement or of bars. The difference between these apparent long bars and real stripes of cirrus can be detected in a moment by turning in any direction. The bars of strato-cumulus follow you by remaining parallel to the horizon whichever way you look, for the linear arrangement is only an effect of perspective; while cirrus-stripes always converge to the same point on the horizon. Fig. 19, which is a

FIG. 19.—Strato-cumulus; roll cumulus.

fair specimen of this kind of cloud, is engraved from a photograph by the author in lat. 18° S., long. 4° E.; that is, in the south-east trade between Goree and Cape Town. We see at once that the sky is too irregular for pure stratus, but that the masses into which the cloud is gathered have nothing in common with pure cumulus; and also very clearly that the linear arrangement increases towards the horizon. This is the cloud to which the term "roll-cumulus" has been unfortunately applied in England.

Though true strato-cumulus is not really allied to cumulus at all, we sometimes see a cloud of this type with a distinct but irregular cumulus form in places. This must also be called strato-cumulus, as it merges by indistinguishable gradations into the purer form of the same name.

Sometimes also we get strato-cumulus from a development of cumulo-cirrus with fine weather in the temperate zone; so that the name and form alone of this cloud tell us little either of its origin or portent.

NIMBUS.

The term nimbus need not detain us long, and then principally to explain the unfortunate confusion which has arisen from the uncertain use of this word.

Every kind of cloud from which rain falls is a nimbus, and there are practically two sorts—cumulo-nimbus, the rocky cumulus-cloud from which rain falls in squalls or showers; and pure nimbus, a flatter cloud, more like heavy strato-cumulus, that forms from or under cirro-stratus in front of extra tropical cyclones. Howard calls nimbus "a cloud, or system of clouds, from which rain is falling. It is a horizontal sheet, above which the cirrus spreads, while the cumulus enters it laterally and from beneath."

Hildebrandson uses the word in a more contracted

signification, and reserves the name of nimbus for the lower layers of dark torn clouds from which rain falls. Poey calls the same broken clouds fracto-cumulus.

Weilbach designates by nimbus the property which a cloud manifests to be or to become a source of rain in particular circumstances, and then gives three varieties—nimbo-pallium, the rain-cloud in front of cyclones, which we have called pure nimbus; nubeculæ, or scud; and nimbo-stratus, the rain-cloud, in rear of cyclones, which we have designated cumulo-nimbus. He also gives a plate marked cumulo-nimbus, which is identical with our application of the same name.

The reason for making nimbus a class of its own comes from the fact that a sudden striking change comes over the look of the upper surface of a cloud the moment rain begins to fall, the precise nature of which we cannot at present explain.

The following remarkable description of the changes which often take place in the appearance of the summit of a cumulus when it commences to discharge rain, is given by Mr. Ley:—

"Under a summer sky a massive cumulus begins to form a few miles distant from the observer. The atmosphere being nearly calm up to the height of twelve or fourteen thousand feet, the cumulus preserves its hemispherical form, and an enormous aggregate of cloud-matter is produced, the contents of which may occupy a space of upwards of a hundred cubic miles, while the extreme opacity of the cloud shows that the water-spherules which compose it are somewhat closely packed. No rain falls from such a cloud while it preserves the hard outline of its upper portions and its general hemispherical figure. Suddenly the summit of this cloud becomes soft-looking, and spreads out laterally in cirriform fibres, this change being always simultaneous with the fall of a sheet of rain out of the cloud.

"The electrical charge which prevented the collision of the particles composing the cloud while these particles remained spherical, has been suddenly diminished in the upper portions of the cloud as soon as these particles are congealed into ice-needles, from the edges and extremities of which the electricity immediately escapes.

"The particles, now only moderately electrified, unite, and in their rapid descent absorb the smaller spherules with which they come in contact.

"The falling rain, and perhaps still more rapidly the disruptive discharges, if such occur, further tend to 'tap' the electricity of the cloud—*i.e.* to lower the potential of the cloud-mass—and the shower-making process continues till all, or nearly all, of the lower portion of the cloud has disappeared. Now, in those instances in which there is not only very little motion in the lower layers of the atmosphere, but also in the higher, the ice-cloud left in the upper regions of the atmosphere is a true cirrus, the curls and twisted forms of which are probably due to slight lateral inequalities of pressure produced by the processes of condensation and congelation. A cirrus so produced may hang nearly motionless for upwards of twenty-four hours in the sky, or may more commonly drift very slowly over districts from which the shower which produced it was invisible."

Mr. Ley further says that he can always tell by the

appearance of the top of a cloud whether it is discharging rain or not; but the cirrification of rain-cumulus is certainly not necessary to precipitation.

The allusion to the discharge of electricity at the moment of precipitation refers to an idea which has much to support it—that free statical electricity tends to keep small condensed globules of vapour apart.

Lord Rayleigh has proved, experimentally, that moderately electrified water-drops tend to coalesce, but that strongly electrified drops repel one another. The precise bearing of this on the formation of rain cannot be given, but it shows unmistakably that there is a real connection between rain and electrical manifestations. We may, however, remark that it is almost certain that the presence of electricity is quite secondary to the other influences which develop rain.

Electricity may determine the precipitation of a cloud, but it cannot give rise to the ascensional current, which is the primary cause. Connected with the appearance of the top of cumuli there is a well-known saying, that "When clouds look woolly, snow may be expected." This refers to the tops of cumuli, and not to the ordinary woolly cirro-cumulus which is so so often seen in summer. There is reason to believe that this woolly look is really due to the cloud being composed of frozen, and not of liquid, globules of water. The author has made some observations on the sudden splash of rain or hail, which often comes directly after a flash of lightning, and the thunder-clap which accompanies them. By measuring the time which elapses between the flash of lightning and both the thunderclap and the splash of rain that follows to one-fifth of a second, he has found that the flash, the clap, and the splash of rain may be supposed really to occur simultaneously, but that the three impressions reach the earth's surface at different times, because light, sound, and a falling body all travel at various rates. Thus light travels practically instantaneously; sound at the rate of about 1100 feet a second; while rain-drops fall a definite distance in any given time, under the influence of gravitation. This would be proved if we found that the distance of the origin of lightning, as measured by the velocity of the sound of the thunder, was the same as that measured by the velocity of falling rain. For instance, on one occasion the interval between the lightning and the thunder was five seconds, while the rain did not arrive for nineteen seconds. Now, calculating the distance of the origin of the lightning from the velocity of sound, we find the altitude to be 5500 feet; while the distance through which a drop would fall in nineteen seconds would have been 5800 feet. The difference is only 300 feet, which is very little considering the nature of the observations, and the unknown retardation of a falling drop from the resistance of the air. In practice the thunder always arrives before the rain; in fact, we may consider that the same disruptive discharge of electricity sends three messages to the earth at different rates, and to different senses—the light to the eye, the sound to the ear, and the rain to the touch.

UNCLASSIFIED CLOUDS.

So far for the great subdivisions of cloud-forms, but we must now mention a few minor forms, because they have some importance in judging weather.

CIRRO-NEBULA, OR CIRRUS-HAZE.

Sometimes, as a cyclone approaches, in any part of the world, and we are very nearly on the line of its path, we see a blue sky first get white, then grey, and then work up to drizzling rain, without the formation of any true cloud-form. When this happens, the sky is said popularly to sicken, and this is an almost infallible sign of rain, and probably of wind. Mr. Ley has proposed the name "cirro-nebula," or "cirrus-haze," for this appearance, and the term seems most appropriate. We may, however, observe here the necessity for our caution about the words *cirro*, *cumulo*, etc., conveying a rough idea of the height of clouds.

This cloud has no fibrous or hairy structure to which the name of *cirro* could be strictly applied; but if we also lay down that the word *cirro* is to convey the idea of a high level cloud, then the word *cirro-nebula* is quite correct. It is invariably formed at a great height, and as it nearly always shows a halo when the sun or moon shines through it, we may assume that it is composed of frozen particles, or ice-dust. After it has formed, we can often watch a layer of cirro-stratus being formed underneath the haze. From all this we may draw the important inference that, though the front of a cyclone is characterized by excessive warmth on the surface, the upper strata are then very cold.

SCUD, WRACK.

Under any mass of cloud which is verging on the precipitation of rain, we have just mentioned that small detached clouds are frequently seen in rapid motion. In England they are called "scud;" in France, *fuyards*, or *diabletons;* while Poey suggests the name of fracto-cumulus. If, instead of being shapeless, they are raggy, they are then known in England as "wrack," from their drawn-out appearance. In all cases it is obvious, from the above description, that they are rather associates of heavy rain-clouds than true prognostics. What we have to explain is their origin. This seems to be simply, that in very disturbed weather small masses of cloud form like ordinary ragged clouds, from the irregular nature of the rising currents, while the apparently very rapid motion comes from their being nearer the surface than ordinary clouds.

CLOUD-WREATHS.

Sometimes, in front of certain kinds of squalls and thunderstorms, we see a long, narrow roll of black cloud moving rapidly, broadside on, and a very well-developed example will be found in Fig. 56, under the heading of "Pamperos." Dark cloud-wreaths in a very much less pronounced form are very common in England before certain classes of squalls and showers, and in front of a curious light grey vault of rain-producing cloud, as illustrated in Fig. 54, in our chapter on Squalls.

There are, of course, many other minute differences of the various classes of cloud, to which it is impossible even to allude in an elementary work like the present. Our object will have been attained if we have succeeded in explaining the general principles of cloud-formation, and the method of making prognostications from their varying appearance. A few hours spent in watching the changing and degrading forms of a sky which is covered by detached cumulus, or the very different modifications almost from minute to minute of cirro-stratus, will better assist any one to understand the nature of cloud-forms than reading pages of the best printed matter.

We may conveniently summarize here the various varieties of cloud-forms which we have already described. The general idea of our classification has been that, though for large bodies of observers all practical men are agreed that eight or ten principal varieties are all that can safely be used, still more advanced cloud-observers will not be satisfied with so coarse a subdivision, and that therefore more minute varieties are necessary.

The ten principal varieties are therefore printed in capitals, while the minor varieties are denoted by smaller letters.

But there is another point in our subdivision of varieties. Almost all the smaller varieties are so rare or transient that for practical purposes they may be neglected; but if, on the contrary, the ten main words are restricted to the forms of clouds we have described under them—that is, *cumulus*, pure rocky cloud; *stratus*, pure sheet cloud; *cirrus*, pure wispy cloud; *cirro-stratus*, thin, high, wispy, or striated sheet cloud of all sorts; *strato-cirrus*, a similar low cloud; *cirro-cumulus*, fleecy cloud at high level; *cumulo-cirrus*, the same, lower down; *strato-cumulus*, extended lumpy cloud; *nimbus*, low rain-cloud; *cumulo-nimbus*, rocky rain-cloud—then the author can say, from an experience of cloud-observation in all longitudes, and in latitudes ranging from 72° north to 55° south, that ninety per cent. of skies in every part of the world can be sufficiently accurately defined by these ten words.

Varieties of Clouds.

With the mean height of the principal varieties at Upsala in summer:—

			Feet.
High	Cirrus		27,000
	Cirrus stripes.		
	Cirrus haze.		
	Cirro-stratus		27,600
	Cirro-cumulus		20,000
Middle ...	Strato-cirrus		15,000
	Cumulo-cirrus		12,000
	Festooned cumulo-cirrus		
	Mackerel sky		
Low	Strato-cumulus		6,000
	Cumulus	(base)	4,000
	Turretted or line cumulus		
	Festooned cumulus		
	Cumulo-nimbus	(base)	4,000
	Cumulo-stratus		
	Nimbus		4,500
	Stratus		1,900
	Scud, wrack		
	Wreaths.		

Of course it will be understood that the levels given here for Upsala do not apply to all the world, but vary with the season and latitude. They are introduced here to illustrate a great principle, which holds from the equator to the pole, that clouds tend to form at a few definite levels, rather widely separated. Thus at Upsala we find the high, middle, and low clouds at about 25,000, 14,000, and 6000 feet respectively.

If we wish to simplify still further cloud names for observers who cannot see the difference between the ten principal varieties, we can put cirro-stratus and strato-cirrus collectively into cirro-stratus; cirro-cumulus, and cumulo-cirrus into cirro-cumulus; and treat cumulo-nimbus simply as cumulus. Then we get only seven terms: cirrus, cumulus, stratus, nimbus, cirro-stratus, cirro-cumulus, and strato-cumulus; and still ninety per cent. of all skies could be defined by these words, only not with the same precision as by the ten varieties before mentioned.

Modern Improvements.

We will conclude this chapter with a few remarks on how far the increased knowledge of clouds improves our capabilities of forecasting, both for a solitary observer, like a man on board ship, or for a central meteorological bureau, which can construct synoptic charts by telegraphic observations from distant points.

No advance is more important than what we have insisted on so much in this chapter—that no sky can be read mechanically, without reference to its surroundings. We have seen that there is fine-weather cumulus as well as cumulo-nimbus, and a fine-weather as well as a dangerous cirrus, while fleecy clouds have not the same import in London as on the equator. In practice the good and bad forms can rarely be mistaken, but sometimes very difficult cases arise. Clouds, in fact, tell us by their appearance, what might be written in words, that more or less damp air is rising or falling under certain conditions of upper and lower wind-currents. The significance must be judged by the surroundings and antecedents, just as the sense of many words can only be judged by the context.

Then the cloud-observer who has added the modern knowledge of the motion of cirrus to the older lore, which is only concerned with the kind of cirrus, would sometimes be able to indicate weather better than his neighbours. And even if both would agree as to the approach of rain, the former would sometimes be able to give much greater precision to his prognostications than the latter.

Another most important advance has been made by Mr. Ley. He finds that, like every other phenomenon of a cyclone, the relation of the upper to surface winds is relative to the direction in which the depression is moving, and that, to a certain extent, the direction of the highest clouds coincides with that of the path of the cyclone. For instance, if the cirri in front of a cyclone come from the south, the depression will probably also advance from that direction at some distance to the west of the observer; while if they come from the west or north-west, the depression will then most likely move from the westwards also at some distance to the north. Unfortunately, the details of these relations are too complicated and too local for an elementary work.

But when we come to think how far cloud-observations may assist a central office, the case is different. The foundation of all modern cloud-knowledge turns round the relation of cloud-forms to the shapes of isobaric lines, so that though, as Mr. Ley says, an isolated observer in the English Midlands can plot out on a map the general distribution of atmospheric pressure, and of weather existing over the whole of the British Isles at the time of his observations, with very considerable accuracy, still an observer at the central bureau could do so with absolute accuracy. He could also often telegraph to an observer in the Midlands, while the sky was still cloudless there, that cirrus of a particular type would form after a certain time.

Where cloud-observers can assist a central office is in forecasting that kind of rain which is associated with the small secondaries and non-isobaric rains that hardly show on synoptic charts. These will be abundantly discussed in our chapter on the subject. For instance, suppose the morning reports give rather ill-defined isobars, with no rain, but a good deal of cloud at various stations, the central forecaster would only say generally fine weather, with, perhaps, local showers. But now, if, instead of telegraphing the vague word "cloud," the observers could not only define the kind of cloud accurately, but also give information as to the direction and velocity of its motion, relative to the surface-wind, then in many cases the central office would see the incipient formation of small rainy secondaries, and the forecasts sent to the different districts would gain much in accuracy and definiteness.

PART II.

ADVANCED.

CHAPTER IV.

ISOBARS.

In our introduction to prognostics, we have already explained the leading features of the science of isobars, and of their relation to the changes in the readings of the meteorological instruments which are most usually observed. In our chapter on Clouds, we have also introduced the reader to the idea that over the surface-circulation of the air round cyclones and anticyclones there is an upper circulation of a very different character. But we must now go more deeply into the subject, so as to explain many details which could not then be conveniently given, and we shall not only complete our description of the nature of cyclones, anticyclones, etc., but also describe the two remaining forms of isobars—V-shaped depressions and cols—which were omitted in the previous chapters.

Cyclones.

We have already sufficiently explained the broad features of a cyclone, and the wind and weather which are associated with it. The reader will now comprehend what is meant by the centre, the trough, the front or rear, the intensity, the path, and the velocity of the cyclone; and he will also understand that it is generally associated with bad weather, and rapid shifts of wind according to very definite laws.

What we want to consider now is the kind of circulation which constitutes a cyclone, and some points connected with the propagation and motion of this particular kind of low pressure. Conventionally, we shall not call a low pressure a cyclone, unless the isobars form a well-defined closed-curve. With the exception of V-shaped depressions, any other irregular area will be called by the generic name of a "depression."

General Circulation.

As the surface-wind in a cyclone is always a little incurved and the upper wind always more or less outcurved, the inference is irresistible that the main body of the air near the centre of a cyclone must be rising; otherwise, as the wind is always blowing in, the cyclone would soon fill up if there was no escape upwards. To this ascensional movement undoubtedly must be attributed the rain and cloud which we find there—rain near the centre, where the ascensional impulse is strongest; cloud round the outside, where the uptake is less strong. From this we can readily understand the effect of what we have called intensity in a cyclone. It is not difficult to conceive a cyclone which possessed so little intensity that it could only develop cloud in the centre. Then, if from any cause the intensity of the ascensional current could be increased, rain would be developed where only cloud had been formed previously. Thus we get hold of the idea, which we shall work out in some detail in future chapters, of the influences which can modify any existing cyclone.

If, for instance, any cause, such as the heating of the ground by the sun, increased the velocity of the wind, and so poured more vapour-laden air into the centre in a given time, then the uptake would be greater, and the tendency to form rain would be increased. Similarly, if the wind was unchanged, but local causes, such as a range of hills, gave the inpouring currents an increased ascensional impulse, then, too, the precipitation of rain would be still further developed.

Axis.

Returning now to our conception of the cyclone as a circulatory system, it is manifest that we may consider the whole as constituting an extremely complicated vortex, something analogous to an eddy of water. There is, however, this difference—that a water-eddy sucks down, while an aerial cyclone draws upwards.

The line along which any particle of air may be supposed to move must not only curve irregularly inwards, but also upward, and finally outwards. As the whole is treated as a whirling system, there must be a line, more or less perpendicular to the earth's surface, round which the air rotates in this complicated manner. This imaginary line is called the axis of the cyclone.

There is much uncertainty as to the nature of the circulation round this axis. Some writers have thought of the axis of a top, and believed that the axis of a cyclone can nutate, always keeping a revolving disc of air perpendicular to itself, so that the cyclone would be pressed down on the ground in the direction towards which the axis inclined, and be lifted off, as it were, on the opposite side. This, they say, would explain the anomalies that are sometimes found both in the position of steepest gradients relative to the centre, and in the variable destructiveness of the wind.

With the same velocity, wind will sometimes unroof houses, at other times do little damage; and they consider that in the first case the direction of the wind is a little upwards, in the latter a little downwards. They believe that this conception of an inclined axis is confirmed by the fact that, in tropical cyclones, the small, clear patch of blue sky in the centre of a cyclone is not always exactly over the point of lowest barometer. They would then consider that the axis of the cyclone is like a telescope pointed upwards, at some angle from the ground, instead of truly vertical.

The insuperable difficulty in the way of all this lies in the fact that a cyclone is often one or two thousand miles across, and certainly not more than ten miles deep; so that the amount of tilt required to give the observed deflection of isobars would be sometimes 20° or more, and that, with a disc of two thousand miles, would be impossible under the conditions of our earth.

If, for instance, we glance back for a moment to our typical cyclone (Fig. 2), we see that the isobar of 29 ins. is not concentric with that of 30 ins. The idea would be that a vertical-axis cyclone would have concentric circular isobars, but that both the oval form and setting back of the isobars at one side are due to the axis being tilted forward towards the word *Front.*

Another theory to account for all these facts supposes that a cyclone is made of a series of flat oval horizontal sections, but that these are not superimposed concentrically one on the top of the other, but pressed successively more or less to one side by surrounding influences. In Fig. 2 this shunt would have been towards the rear, instead of forwards as by the preceding hypothesis.

This view is probably partially correct, though it is impossible to suppose that the air does not get more or less inclined upwards at times, for no cyclone is ever absolutely symmetrical.

We often see the conical revolving cloud of a whirlwind or tornado bending about like the trunk of an elephant, with both a true axial inclination and a certain amount of sideways shunt. Here, however, the vertical height is enormously greater than the diameter, which is just the opposite to the proportions of a cyclone. Anyhow, if we suppose that upper winds follow the same laws as surface-currents with relation to isobars, observations on cirrus-clouds tend to the belief that the axis of a cyclone is very often inclined backwards from the direction in which the cyclone is moving, as if the surface portion was going faster than the upper.

This is just the converse of what might have been expected *à priori*, that surface friction would retard the lower portions, so that the axis of the cyclone would have been inclined forwards. The whole question is, however, still very obscure.

PROPAGATION.

When we come to consider the nature of the propagation of a cyclone, we are met with many difficulties. At first it might be thought that a cyclone could be treated as a rotating disc, which was impelled along the earth's surface by some force; but there are conclusive reasons against such a supposition. If this was really the case, we ought to be able to compound the rotation and translation motions of any particle of air in the usual manner; but when we do so, we find that we get winds very different from what are actually observed. Take that portion of the front of a cyclone where the wind from rotation would be south, and suppose it to be compounded with even a slow motion of translation towards the west, then the resulting wind must have a slant from the west —that is to say, it would blow outwards in front of the centre. Now, this is exactly what it does not do. Observation shows that the wind is more incurved in front of a cyclone than in any other portion, and therefore the idea of a rotating disc cannot be maintained. We are almost compelled to believe that a cyclone-vortex is propagated in a manner somewhat analogous to a wave of water. When a wave approaches the shore, the first impulse is always an indraught, though, of course, the motion of the wave is forwards; and when a cyclone approaches, the first impulse is likewise inwards. Here, however, the analogy probably ceases.

Another analogy to vortex-motion is found in the manner in which a cyclone, as a whole, is deflected by areas of high pressure.

We know by experiment that a vortex-ring of smoke has great stability, and can be twisted and deflected like an elastic body. Similarly we see a cyclone move up against an area of high pressure and be rebuffed from it, even though the cyclone may be one thousand or more miles in diameter. Of course, it must be understood that in a vortex-ring each particle of air revolves in a complete circle, while in a cyclone any particle hardly describes a semicircle, so that the analogy is only very partial.

Stability.

However, from this conception of a stable vortex we can understand what has long been a puzzle to meteorologists—why great changes of temperature between night and day can be associated with fine weather.

Differences in temperature between two adjacent areas have always been supposed to set up currents of air, and from this it has been thought that weather-changes could be deduced. In practice, however, it is far otherwise. In some of the most settled climates in the world, such as Persia or Northern Africa, the difference between the day and night temperatures is often from 30° to 40° Fahr.; while in unsettled climates, like Great Britain, 10° Fahr. (5° C.) is a large amount.

Then consider that a difference of 16° Fahr. (9° C.) makes as much change in the density or specific gravity of a cubic foot of air as a change of one inch of mercurial pressure, and we may well wonder why such great changes have so little effect on weather.

But when we know that a cyclone and, as we shall

afterwards see, an anticyclone are both to a certain extent vortices, and that all the world is generally covered by one or the other, then we can readily understand that as such they have great stability, and that, though changes of temperature might affect the velocity or direction of the general circulation, they could not break up or destroy any existing system. In fact, the atmosphere is not an inert mass, ready to be swayed by any trifling disturbance, but is always broken up into circulatory systems, which possess a very considerable amount of stability. When we come to discuss diurnal variations of meteorological elements, we shall find that the difference of day and night temperatures only impose a very small modification on the general character of the weather

Influence of Rainfall and Temperature.

There are two other points connected with the origin and motion of cyclones which need only be alluded to in an elementary work like the present—the influence of rainfall, and the surrounding distribution of temperature. In most cases the greater portion of the rain falls in front of the cyclone's centre, or trough, and it used to be thought that, the rain being produced by some unknown cause, the cyclone ran after the vacuum which was left by the condensation of vapour. It was, moreover, believed by some that this condensation of vapour was also the origin of the cyclone. Air was supposed to rush into the vacuum, to pick up circulation through the influence of the earth's rotation, and thus to form an eddy.

Later observations have, however, completely disproved

both these views. Under certain conditions of surrounding pressure, the greatest rainfall in a cyclone occurs in rear of the centre; and the commencement of a cyclone is neither always preceded by rain, nor does the depth of a cyclone bear any relation whatever to the amount of precipitation.

In practice the heaviest rain is with the slight gradients of a secondary, and in very heavy showers the barometer generally rises.

The general view we wish to present is that of the air rising over a hot equator, and pouring down towards the pole by its own weight. Mere irregularities of this downflow would be sufficient to form eddies without any condensation of vapour to cause vacuum, and develop heat, as we can readily prove by experimenting with water or smoke.

If the atmosphere was absolutely vapourless, and the sun moved round a motionless earth, the irregular overflow of air would form eddies of some sort, very different from those we now know. If the same vapourless atmosphere rotated with the earth, round a stationary sun, as at present, cloudless cyclones and anticyclones would undoubtedly form, perhaps not so very different from those we know now; while the addition of water-vapour would make the rotational systems precisely such as we now observe.

Though heat and vapour do play a considerable part in the mechanism of a cyclone, we shall abundantly show, in the course of this work, that cyclones are only incidents, as it were, in greater movements of the atmosphere, and that most temperature-changes are due to differences of radiation, caused by the wind, calm, cloud, or blue sky associated with different kinds of aërial eddies.

An enormous amount has been written in Germany, India, and the United States on the influence of surrounding temperature, and of the latent heat released by the condensation of vapour into rain in determining the course of a cyclone. Most of this is, however, too much mixed up with theory, and too much encumbered with mathematical formulæ, to find a place in this work. There is, unfortunately, no consecutive account of those investigations in the English language, except certain papers of Professor Ferrel, published by the Signal Office in Washington, which cannot be purchased; but they are admirably given in Sprung, "Lehrbuck der Meteorologie," and J. von Bebber, "Handbuch der ausubenden Witterungskunde."

We shall recur to this subject in our chapter on Forecasting by Synoptic Charts, as by that time we shall be better able to understand the nature and surroundings of a cyclone.

Our knowledge of the nature of cyclones would be much more complete if we knew how high in the atmosphere they extended. In thunderstorms we often get a complete circulation of the wind on the surface, while the upper currents retain their direction unchanged, and the drift of the storm is usually with this higher wind. Unfortunately, we cannot trace a continuous development of these small circulations into regular cyclones.

Tropical and Extra-Tropical Cyclones.

It has been too much the custom in meteorological books to treat tropical cyclones apart from similar disturbances in extra-tropical or temperate regions. We have made numerous researches on the subject in India, the China Seas, Japan, and Mauritius, and found that, though the general character of all cyclones is the same, there are differences of detail which throw an immense amount of light on the cause of the great variety in the appearance of the sky in different parts of the same cyclone.

All cyclones agree in the great features of the wind rotating round the centre with a variable indraught, and of an upward and outward circulation of the higher currents.

No more conclusive proof of this can be found than the fact that cyclones often pass out of the tropics, and then join or coalesce with others which have been formed without the tropics. Two similar eddies can easily unite, but two that rotated on different systems would infallibly break each other up.

The typical cyclone in all parts of the world is certainly oval, with the inner isobars usually closer to the rear than to the front; and the rain extends further before than behind the trough. But the tropical cyclone has a striking feature which is absent in our latitudes. There is a patch of blue sky over the calm centre, which is well known in hurricane countries as the "eye of the storm," or as a "bull's-eye." Then cirrus and halo appear all round a tropical cyclone, while they are never seen in rear of a European storm; and though the way in which the rain seems to grow out of the air in front of a cyclone is the same everywhere, the sky and clouds in rear of a hurricane are much softer and dirtier than in temperate cyclones. There is not that sharp difference between the quality of clouds in front and rear which is so striking in higher latitudes. Still greater is the absence of any marked squall or change of weather during the passage of the trough in the tropics—that is, at the moment when the barometer begins to turn upwards. Some who study hurricanes have scarcely noticed any change then; and all are agreed that the trough-phenomena are very slight.

We have already shown, in our chapter on Prognostics, that a cyclone has, as it were, a double symmetry. One set of phenomena, such as wind, cloud, and rain, are grouped round the centre; while the second set, such as the different character of the heat and clouds in front and rear, and the line of squalls along the line where the barometer begins to rise, are related to the trough of the cyclone. If we call the first set the rotational, and the second set the translational, phenomena of a cyclone, we find that the former are all more marked in the tropical, and the latter in extra-tropical, cyclones. Then, if we examine the charts of cyclones, we see that, while tropical hurricanes are much smaller, and have much stronger winds than any others, they only move from two to ten miles an hour; while extra-tropical cyclones rotate much more slowly, but are propagated at a rate of from twenty to seventy miles an hour.

Thus we might readily suppose that what we call rotational phenomena are really due to the circulation, and the translational phenomena to the forward motion, of a cyclone; and we are confirmed in this view by an examination of Japanese typhons. That semi-tropical country is traversed by cyclones of two different types at different seasons of the year, that move with different velocities, and they find that all the trough-phenomena are more marked in the quickly moving cyclones than in those whose progress is slower.

These researches also lead to another most important conclusion—that the character of cloud and weather depends on the position relative to the front of a cyclone, and not on the direction of the wind. Cyclones in Europe move towards the east, and the dirty sky comes with a south-east wind; while in the northern tropics hurricanes move towards the west, and the same sky comes with a north-west wind. People sometimes say that of course the rear of a cyclone must be clear, because of a cold, dry north-west wind; but when a cyclone moves west, even in Europe, that wind becomes close and dirty.

We shall defer our consideration of cyclones in the southern hemisphere till our chapter on Winds, because it is only the direction of the wind, and not the sequence of weather, which is altered in comparison with northern storm-systems.

Anticyclones.

If we turn to the diagram (Fig. 16) of surface and upper winds in an anticyclone which we gave in our chapter on Clouds, we shall see at once that it presents some analogies, as well as some very striking contrasts, to the cyclone figure. The anticyclone blows round and out below, round and in above, and therefore the conclusion is obvious that the air in the centre of an anticyclone must be descending. It must, then, necessarily be unusually dry, and this is just what observation shows it is. Then exactly the same argument holds as in a cyclone, that, as a whole, an anticyclone is a complex vortical system which possesses so much stability that great diurnal changes of temperature do not affect it as a whole.

It may be well to note the higher character of the explanation of weather which we can give now, as compared to what we said when treating of prognostics. Then we merely said that, as a matter of blind observation, the centre of a cyclone was rainy, and that of an anticyclone bright. Now we show that these two varieties of weather are the necessary product of different kinds of atmospheric eddies.

Pressure over Cyclones and Anticyclones.

Simultaneous observations at the top and bottom of high mountains have demonstrated that the difference of pressure for a given height is always less in cyclones than in anticyclones; also that the fall of the barometer is always less pronounced on the summit than at the base of a mountain. For instance, if the difference of pressure between a high and low level station was four inches in a cyclone, it might be four and a quarter in an anticyclone; and if the barometer fell an inch at the sea-level, the fall might only be about eight-tenths of an inch on the top of a mountain five thousand feet high.

The inference which is drawn from this is that, as we ascend, the gradients between a cyclone and its adjacent anticyclone must diminish, and it is by no means improbable that if we went up high enough we should find them inverted; that is to say, that the higher pressure would be over the cyclone.

In Fig. 20 we have drawn an ideal sketch of the

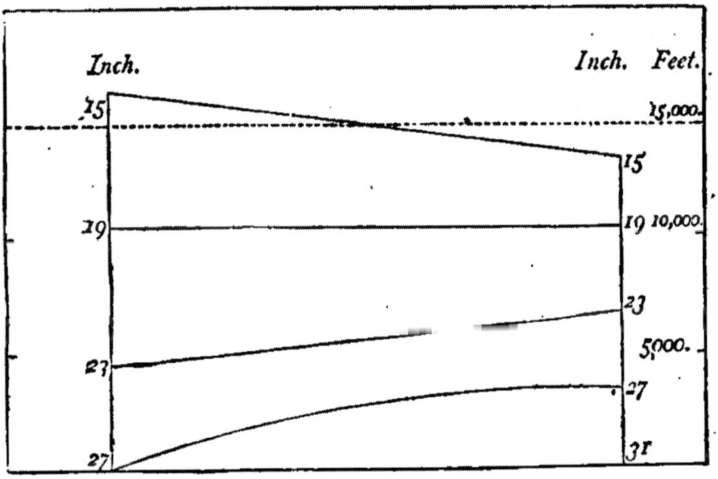

FIG. 20.—Probable vertical gradients over cyclone and anticyclone.

probable so-called vertical gradients over a cyclone and its adjacent anticyclone. The line at the bottom represents the level of the earth's surface. If the pressure is 31 ins. over the anticyclone, and only 27 ins. over the cyclone, the vertical isobar of 27 ins. must be as we have drawn it. But, as pressure decreases more rapidly over an anticyclone, it seems probable that at a certain level—

which we have assumed here as ten thousand feet—there would be no gradient either way; and that still higher up the pressure would be actually greater over the cyclone, and the gradient inverted as compared to that on the surface. It is evident, therefore, that, though we may use the analogy of cyclones to whirlpools, we must not picture the former to ourselves as saucer-shaped depressions of the whole envelope of our atmosphere, like the little eddies that pit the surface of a flowing river. The important bearing of these vertical gradients on the problem of measuring heights by the barometer will be very obvious.

This brings us to a curious question, as to what gradients give direction to the upper winds. Some have maintained that the uprushing currents of a cyclone have so much momentum that they can override moderate gradients. Others, on the contrary, hold that it is only by a complete or partial inversion at high levels of the gradients which are found on the surface, that the observed phenomena of upper currents can be produced; but materials do not at present exist which can decide the question, and some of the published sketches of the higher isobars over a cyclone are more than problematical.

Another point on which considerable uncertainty exists is the relation of anticyclones to cyclones. There is no doubt of their partial dependence on one another, for cyclones always tend to travel round the anticyclone on whose edge they have been formed. But, on the other hand, we sometimes find cyclones which have detached themselves from their generating anticyclones, and this seems to prove that they can exist independently.

Antithesis of Cyclonic and Anticyclonic Weather.

Perhaps the best method of showing the antithesis between cyclone and anticyclone weather will be the method we have adopted in Figs. 21 and 22. We all know how much weather is affected by the time of day, as well as by the season of the year, and by local peculiarities. We have, therefore, selected two charts for the same day of different years, at the same hour of the morning, and for the same portion of Western Europe. Every diurnal, seasonal, or local influence is therefore identical in both cases, and the whole of the difference of wind and weather which we find between the two days is entirely due to cyclonic or anticyclonic influences.

In Fig. 21 we give the synoptic conditions of pressure, temperature, wind, and weather over Western Europe at 8 a.m., May 17, 1877. There we see a small oval cyclone of very moderate intensity lying over the south-west of England. Round this the wind circulates in the usual manner, but, as the gradients are not steep, the force nowhere exceeds a fresh breeze, as at Brest. Near the centre, and some distance in front, we find, by looking at the weather-symbols, nothing but rain reported; outside the rain, an overcast sky or detached clouds; and beyond them, blue sky in a few places. The path of the cyclone is marked by the letters *x*, *x*, so as to give the position of the front. Lastly, the isotherm of 60° Fahr. (16° C.) runs just north of the Pyrenees, while that of 50° Fahr. (10° C.) stretches from the north of Scotland to Denmark.

Now turn to Fig. 22, where we give the same data for May 17, 1874, at the same hour. Then an anticyclone lay over the British Islands; the gradients were much less steep, and the wind, therefore, was everywhere light and variable. For this reason the general circulation is not so marked as in the preceding chart, but still it is very evident that on the whole the wind blew round and out in the direction of the watch-hands. Then the weather-symbols are very interesting. Almost every station which

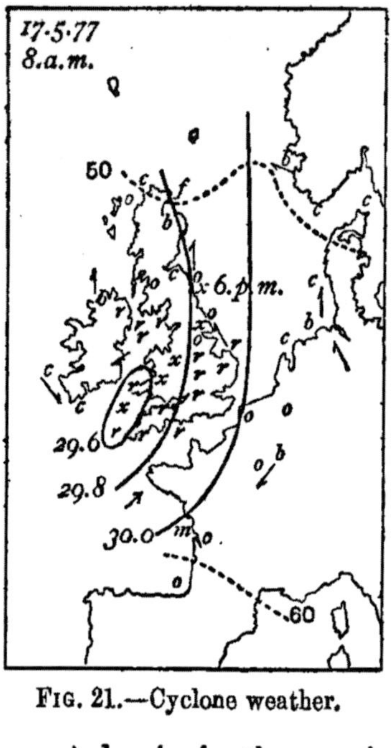

FIG. 21.—Cyclone weather.

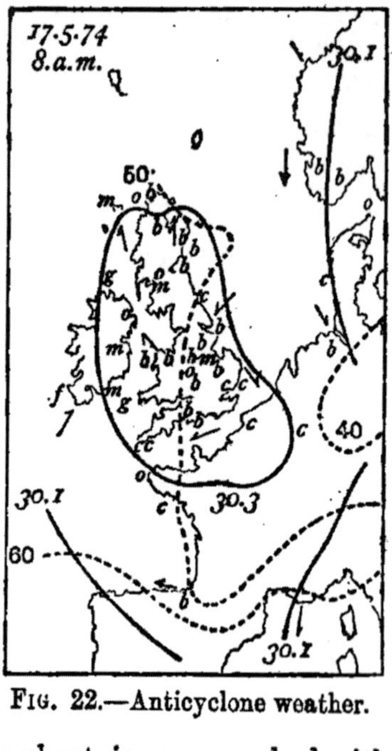

FIG. 22.—Anticyclone weather.

reported rain in the previous chart is now marked with *b* for blue sky, or with *m* for radiation mist in several places. Then, though, as at Fano, in Denmark, we find the same symbol, *c*, for detached cloud in both maps, we know that it does not refer to the same kind of cloud. The temperature also shows a marked contrast. The isotherm of 60° has not much altered its position, but that of 50° Fahr. (10° C.) bends abruptly south from the north of Scotland, across England, and the west of France, while temperatures below 40° (5° C.) are reported from Hanover and the Netherlands. This is partly due to the prevailing set of the wind being from the north, whereas in the preceding chart it was from the south or south-west.

From these examples, we see that the whole of the difference of weather on the two days was produced by the difference of isobars, which we may put thus:

The weather was wet on May 17, 1877, over England, because then the atmosphere was eddying in that manner which we call cyclonic; while it was fine on May 17, 1874, because then the eddy-circulation took the form which we call anticyclonic, and drew down dry air from the upper regions of the atmosphere.

V-Shaped Depressions.

We must now describe a very interesting shape of isobars, to which the name of V-depressions is applied in England, but which the German writers call "tongue-formed" depressions. In these the isobars are shaped like the letter V, and enclose an area of low pressure. In the northern hemisphere the point of the V is usually directed towards the south, as in Fig. 23. The wind follows the universal law of gradients; being from south to south-west in front, and from west to north-west in rear of the trough. This latter line is given at once by joining the southern points of each successive isobar, and in practice is nearly always curved, the convexity being turned towards the east, as in the diagram. As the V is usually moving towards the east, this line marks out the position of all the places at which the barometer, having fallen more or less, has just turned to rise, and is called the "trough" of the V.

These features are common to all V's, but the position of rain divides these depressions into two distinct types. In the first, and by far the commoner kind in Great Britain, a narrow strip of cloud precedes an area of rain, shaped like a portion of a crescent. This is shown in Fig. 23, where the single shading marks the position and shape of the cloud-area, and the double shading that of the rain. The rear of the rain-area is very sharply defined by the line of the trough, which also marks the position of a line of squalls. Beyond this we find detached clouds, and then blue sky.

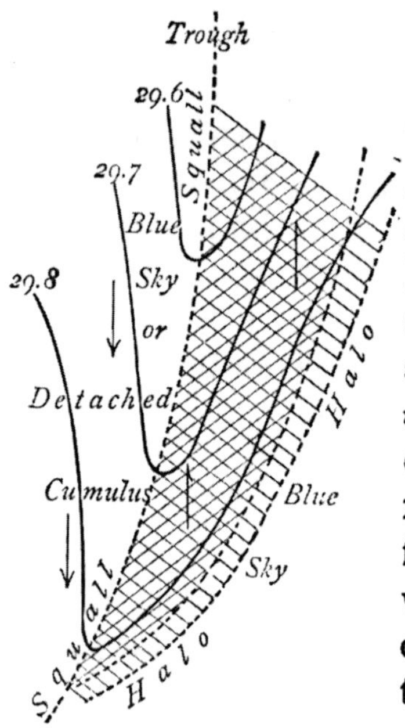

FIG. 23.—Weather in V-depression.

The sequence of weather, as a V of this type drifts over an observer, is obviously from blue sky to halo, cloud, and, later on, rain, with a falling barometer and south-west wind; then a heavy squall, during which the wind jumps (does not veer) to north-west, and the sky rapidly clears as the barometer rises.

In the other kind, which is less common, the front of the V is cloudy, but half a crescent-shaped area of rain is formed in the rear. The front of this area is sharply defined by the trough, while the rear tails off through cloud to blue sky. An illustration of this type will be found in Figs. 49, 50, and 51, in our chapter on Squalls.

The sequence of weather to a solitary observer will be a falling barometer, with a cloudy sky and wind from the south-west. A heavy bank of cloud approaches from the north-west; this passes over him with a heavy squall, the wind jumps to north-west, and the mercury turns upwards. After the first violence of the squall is over, driving rain continues for some time, and dies away gently as the sky becomes clear again.

This class of V is usually followed by a second depression of some sort, as will be seen by Fig. 49 in our chapter on Squalls. There seems to be some connection between the two depressions and the area of high pressure between them, but the subject has not yet been worked out.

V's are generally formed either along the southern prolongation of the trough of a cyclone, or else in the *col* between two adjacent anticyclones; and a V of the first type bears the same relation to a wedge that a cyclone does to an anticyclone; that is to say, one is as nearly as possible the converse of the other. The most interesting feature about V's is their relation to cyclones; for, while a cyclone has, as it were, a double symmetry—namely, one set of phenomena, such as temperature and the kind of cloud symmetrically disposed in front and rear of the trough, and another set, such as wind and rain, symmetrically arranged round the centre—a V-shaped depression has only one line of symmetry—the trough—to the front and rear of which alone both wind and weather are related. When we consider that in extra-tropical cyclones, though the isobars are circular, the wind makes a sudden shift as the trough passes, and that V's have nothing cyclonic at all about them, we can readily understand the difficulties which many felt in accepting the theory of cyclones, and how at first sight it appears much simpler to assume the conflict of two opposing currents, rather than the circulation of a great eddy.

Though a cyclone is a eddy, the opposite properties of its front and rear do not suggest true circulation, while a V is something of a totally different nature. The error consisted in assuming that the bad weather of the V was of the same nature as that in a true cyclone. There is very great difficulty in forming a rational conception of the general circulation in a V. The currents in front and rear of the trough are not exactly opposite, and we know nothing of the motion of the upper currents, so that we must await the results of future research.

Southerly Bursters.

V-depressions are not at all common in the tropics, though we have observed a squall of this type near New Caledonia, in about lat. 22° south ; but we have discovered that a great many of the so-called "southerly bursters" off Australia are due to the class of V's in which rain falls in rear of the trough.

The point of a V in the southern hemisphere is pointed towards the north, while the wind is north-east or north in front, and south-west or south in rear of the trough (see Fig. 39, p. 199).

The first blows in Australia across a burning continent of hot sand, and as the gusts increase near the trough, clouds of suffocating dust are added to the furnace-like heat till existence is almost unbearable. Then suddenly the wind jumps round to south-west or south, and heavy rain is driven before the chilly blasts from the ice-bound southern pole. Temperature occasionally falls 30° or 40° in a single hour. Sometimes the "burster" is associated with the very similar sequence of weather during the passage of the trough of a cyclone.

Cols.

The last shape of isobars which we have to describe is the "col," or neck of low pressure, which lies between two adjacent anticyclones. We will describe a specimen of the most common case of European col. We find in Fig. 24 that a portion of one anticyclone lies over the Bay of Biscay, while another lies to the north-east of this over the Scandinavian peninsula. Then, while one area of comparatively low pressure lies to the north-west of Iceland, another covers Central Europe and the north of Italy, so that a saddle-back area, or neck, of low pressure is found over England. In the middle of the col there is no gradient, and therefore a calm, while all round the winds and weather conform to the usual law of isobars. The weather is dull, gloomy, and stagnant, while in summer violent thunderstorms are frequently found in different portions of a col.

As a whole no sequence of weather can be assigned to a col. It does not move itself, but no law can be laid down to say whether the col will remain stationary, or whether the area which it covers to-day will be occupied by some other type of pressure to-morrow.

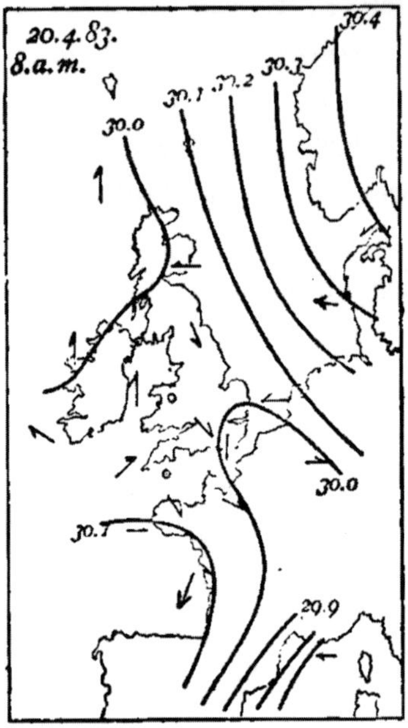

Fig. 24.—Wind in a "col."

The importance of this shape of isobars in forecasting arises from the fact that, as both the anticyclones are usually stationary, the col represents, as it were, a line of weakness, along which disturbances will be propagated.

Unfortunately, though a col can be safely forecast in general terms as for unsettled weather without much wind, the motion of cyclones as they meet a col is most uncertain. Sometimes they pass in a south-eastern direction across Europe between the two anticyclones, while more frequently the main body of the cyclone is deflected or dies out, while an irregular secondary pushes its way more or less across the col.

Origin of Isobars.

Secondary cyclones, wedges, and straight isobars have already been sufficiently described, so that we may now conclude with a few general remarks on the whole question of the shapes or forms of isobars. In the first place, what is the meaning of the seven fundamental shapes? From the analogy of water, to which we have so often referred, there can scarcely be any doubt that just as circulating water can only take a certain limited number of forms, eddies, backwaters, ripples, etc., so air can only move in a limited number of ways, of which these seven are the most important.

A cyclone may be anything from two thousand to fifty miles across; a wedge may fill the whole North Atlantic, or it may also be measured by single miles, but their respective characters are in no way changed. Then, though we are only at present able to give the general nature of the upper and lower circulation in cyclones and anticyclones, there is no doubt that the other five shapes are capable of being worked out in a similar manner. The conclusion is irresistible. Just as we proved that not only the broad features of the weather, but also the minute characteristics of every cloud in cyclones and anticyclones are the product of the nature of their respective circulations, so must the weather and cloud in the remaining five shapes be also the product of some form of atmospheric motion.

In an eddying river the general cause of all motion is the downward current from the source to the mouth; but water cannot slide like a weight on an inclined plane, without forming horizontal or vertical whirls. In the atmosphere the prime source of all motion is the general circulation of air set up between the hot equator and the cold poles. This, like water, must form vortices, or ripples, and isobars map out the varying pressure on the earth's surface induced by the uneven flow of air. If we had a series of barometers at the bottom of a river, and observed them all simultaneously, we should find a defect of weight under the eddies, with an excess of pressure under the backwater, and we might draw isobaric lines which would map out the position of these vortices exactly as we do in air.

But it must be specially noted that the depression of a cyclonic eddy is not entirely due to centrifugal force like that of a water eddy. The depression of an ordinary cyclone could not be produced by the centrifugal force of the fiercest hurricane that ever blew; and the cause of the diminution of pressure in a cyclone is at present unknown.

When we discuss the changes shown on two different synoptic charts, we shall talk of changes in the shapes and positions of isobars as if those lines were graphical abstractions. What we really do is to trace by their means the ever-changing eddies of the atmosphere—the death of old ones, the birth of new ones, or the fusion of several existing systems into a new disturbance of a different type.

To avoid all danger of theoretical errors, we simply define the kinds of eddies by certain abstract shapes of isobars, and then, entirely from observation, we collate certain definite kinds of weather and sky, with different portions of each isobaric configuration.

CHAPTER V.

BAROGRAMS, THERMOGRAMS, METEOGRAMS.

WE have already seen that a series of synoptic charts are, as it were, a series of bird's-eye views, not only of the appearance of the sky, but also of instrumental readings over a considerable area. Owing to the expense of telegraphy, these can rarely be taken oftener than every eight hours, but we all know by experience that very great changes of weather may occur within that time. Every well-equipped observatory is therefore supplied with instruments which record automatically the height of the barometer, of the thermometer, the direction and velocity of the wind, the occurrence and quantity of rain, while a wet-bulb thermometer for moisture and an electrometer are sometimes added.

The trace marked on paper by a barograph is called a barogram; that by a thermograph, a thermogram; both or either of the records left by the wind-instruments are called anemograms; and if all or several of these are combined in one diagram, the whole is called a meteogram, because it is a writing of meteorological instruments. Fig. 25 is a copy of one of the meteograms which are published by the British Meteorological Office, and we propose to devote several paragraphs to its consideration, so as to explain how, by means of these continuous records, we can fill up the gaps in the history of weather-changes

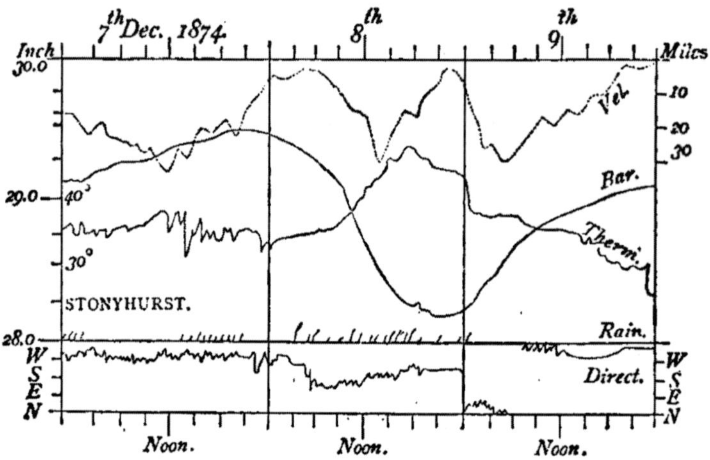

Fig. 25.—A meteogram.

which are shown at considerable intervals only, on synoptic charts.

It should be remarked that in treating of these we no longer deal with generalities, but with the actual changes and variations of each day. Also that the aim and object of all meteorology is to explain the details of the sequence of weather as it occurs at any one place, if possible with great minuteness, and that synoptic charts are only a means to that end. Ever since the discovery of the barometer and thermometer, innumerable attempts have been made to discover the nature of weather-changes by calculations, which were based on the fluctuations of these instruments as observed at any one place, but with very moderate success. We shall very soon see why this was so; and, in fact, why it could not be otherwise, from the nature of things.

METEOGRAMS.

Now for our actual meteogram in Fig. 25. The barometer-trace is marked "bar," with an appropriate scale at the edge. Temperature is marked "therm;" while the upper curve of all, marked "vel," gives the varying velocity of the wind, only we must note that for convenience its base is taken from the top of the diagram, and that, therefore, it is measured downwards. Rain is marked near the bottom of the diagram by oblique lines, which are proportional to the amount of rain measured at the end of each hour, so that we can see at a glance when rain fell and how much of it. Below all is the trace marked "direct," which refers to the direction of the wind. The notation requires a word of explanation, as we have to represent an angular scale on a rectangular diagram. It will be seen that the bottom of the diagram is marked N for north; then as we rise upwards we get successively east, south, west, and back to north again at the top of the direction-figure. By this means we can see at a glance whether the wind is veering or backing. When the wind veers, the trace moves upwards; while, when the wind backs, the curve descends. A little practice is necessary to read this easily, but when once learnt it is very convenient.

But to understand this meteogram fully, we must look at the charts given in Figs. 26 and 27, for the 8th and 9th of December, 1874, respectively. The broad features are very simple. The cyclone which we see approaching in the first chart passes across England to Holland, where we find it in the second chart; and now let us see how

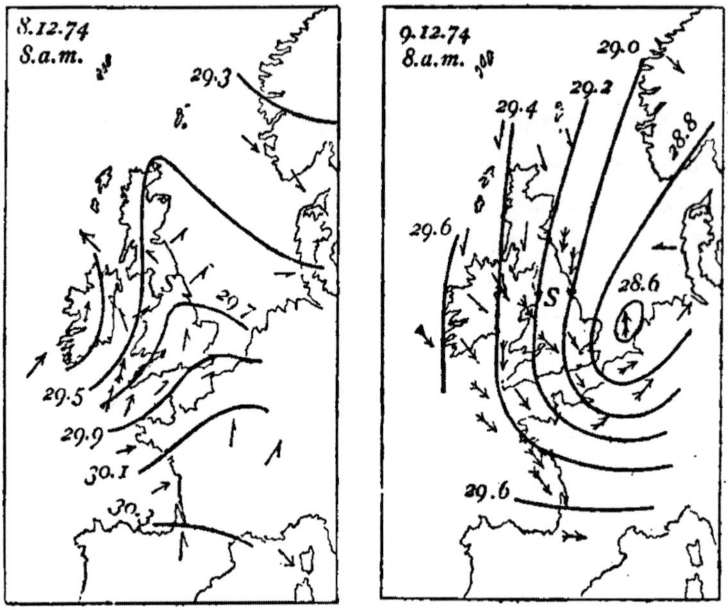

FIGS. 26 and 27.—Charts to illustrate meteogram.

this affected the sequence of weather at Stonyhurst, near Manchester, marked S on the charts.

The meteogram, Fig. 25, refers to the three days, December 7 to 9, 1874, while the charts refer to the two latter days only. The barometer rose till nearly midnight on the 7th, under the action of the wedge which we find over the North Sea on the chart for the morning of the 8th. Then we see by the trace that the mercury was falling fast, owing to the approach of a cyclone. This fall continued till about midnight, when a rapid rise commenced, which lasted all the next day. The charts alone would scarcely enable us to mark the exact line of the passage of the cyclone, but from the wind-traces we know that the centre passed very nearly exactly over Stonyhurst.

The temperature-curve is much more complicated. On the first day we see a most irregular curve, with little trace of the ordinary diurnal range of temperature, for the highest and lowest points were reached within about an hour of each other, and in the middle of the day. This can be easily explained. The station was the whole day under the influence of the rear of a cyclone, with a north-west wind and cold showers. These and driving clouds gave rise to the sudden changes which are marked on the curve. Next day there are more signs of diurnal variation, but the highest temperature was at 6 p.m., and the second midnight is much hotter than the first. Also, though rain was collected almost every hour, we do not see the sudden changes of the preceding day. The interpretation of all this is as follows:—

Commencing from the early morning, the wind began to back from west-north-west to south and south-south-east. This increased the general warmth at the second midnight, and the diurnal range took its usual course on the top of the more general change. Why the rain did not send down the thermometer will be obvious when we know that this all happened in the front of an intense cyclone, and we remember that the soft, warm, drizzling rain which falls there is not like the chilly showers of the rear of a depression.

On the third day the thermogram is still more curious.

Immediately after midnight the thermometer fell 5° Fahr., just as the trough of the cyclone passed, and the wind jumped from south-west to north. During the whole day the temperature fell, and while the greatest heat was at the first, the greatest cold was at the second, midnight. No rain was reported, so the small sudden changes must be the effect of passing clouds. Here it may be well to note that a passing cloud by day will send down the thermometer by hiding the sun, while by night the same cloud will raise temperature by cutting off the radiation of the cold sky. The general explanation is that cold winds set in when the barometer turned, and that their influence apparently completely overrode diurnal variations.

Though the thermometer fell all day—that is to say, there was no hour at which the temperature was not lower than at the preceding one—it does not follow that there was no diurnal range with a regular maximum and minimum; but, as the explanation of the superposition of curves requires some collateral details, we will defer our remarks till we have considered the wind-traces of our meteogram.

First, then, for the velocity-trace. Presuming that the normal course of the wind is to increase regularly in velocity from 4 a.m. till 2 p.m., and then fall gradually again, the most obvious feature in our wind-diagram is the increase of the wind towards the middle of the day; that is to say, the regularity of this diurnal variation in spite of the great cyclonic changes which were going on. The influence of these latter is seen in the comparatively high velocities—from thirty to forty miles an hour—which the wind attained a little after noon on the first two days; but the trace for the last day, December 9, is very different. On that day the strongest winds were before 6 a.m., and during the hottest time of the day, when the wind is usually strongest, the velocities decreased steadily. The reason of all this was that, in the early morning, the steep gradients in rear of the cyclone were passing over the station, while in the middle of the day the gradients were becoming so much less steep that the diminishing velocity due to them entirely overrode any increase which would naturally have occurred from diurnal influences. The calm just before midnight on the 8th, and the rapid rise of the wind to thirty miles an hour directly afterwards, are very interesting; for the calm is that in the centre of an intense cyclone, and the high wind is associated with the steep gradients which we see in rear of the depression in Fig. 27.

The direction-changes for these three days are tolerably simple. The natural diurnal variation of the wind is to veer a little in the forenoon, and back again as the sun goes down; but almost all trace of this is lost in the stormy weather to which our diagram refers. On the 7th the wind kept pretty steady between west and north-west, and there is little sign of diurnal variation. Next day, the 8th, as the cyclone approached, the wind backed rapidly as far as south-south-east; then veered rapidly to south-west; and just at midnight, when the barometer turned, jumped up to the north and north-east; and then backed during the day to about north-west. From these changes, and the sudden jump of the wind from west to north, it is evident that the cyclone's centre passed very nearly over the station. All traces of diurnal variation are, of course, entirely marked by these greater changes.

SUPERIMPOSITION OF VARIATIONS ON CURVES.

The simple conception that the actual weather is the balance or sum of various influences will now be sufficiently obvious, but we must go a little more into detail of all that can be learnt from an inspection of instrumental curves. As the idea of superimposing curves of different kinds of variation one on the top of another, and of so deducing a resulting curve, may not be familiar to some of our readers, we will commence by an easy example.

Here, and all through this work, we shall use conventionally the word "changes" to denote alterations in weather due to cyclones, etc.; and "variations" to denote alterations due to the time of day. The passage of a cyclone, or its replacement by an anticyclone, really changes the weather; diurnal influences only impose a certain variation on these greater changes.

For instance, let us try and find out what would be the nature of the curve left by a thermograph if a regular diurnal variation of temperature, which was highest at 2 p.m. and lowest at 4 p.m., was superimposed on a steady general fall of temperature, due to other than diurnal influences—say, the setting in of cold northerly winds. The line on which the diurnal curve is superimposed is called the level of variation. In the familiar curve of mean diurnal variation, such as B or C in Fig. 28, the straight, horizontal line A represents the mean temperature of the place, and the curves B or C are the resulting traces. If they are unaltered, the only effect of any change in the level of A would be to bring the whole nearer or further from the base of the figure. For the level of A might either be 40° or 70°, but the shape and magnitudes of the diurnal curves B or C would be the same.

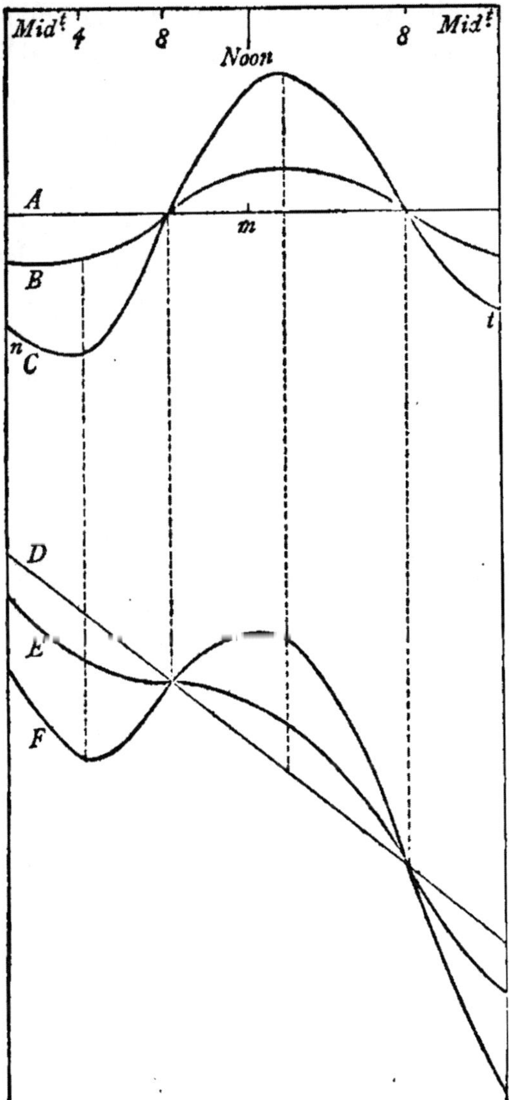

FIG. 28.—Superimposition of curves.

When we come to deal with the significance of a curve for any particular day, the horizontal line A no longer represents the mean temperature of the station, but the level of temperature from general causes independent of the time of day. If the thermograph gave such a trace as B or C, it is possible, from geometrical considerations, to draw the line A, which denotes the level of general temperature on which the diurnal curves are superimposed. The method is as follows. As B and C are both diurnal curves, which only differ in magnitude, but not in character, we will confine our attention to C only. Then, if the instrumental trace gave C, find from it the mean temperature of the day by taking the mean of the readings at every hour; mark this value *m* on the vertical hour-line for noon; join *n* and *t*, the points where the trace cuts the first and second midnight hour-lines; then a line A drawn through *m* parallel to *nt* is the level of variation required. The line *nt* is omitted in the diagram for the sake of greater clearness.

But now suppose that the diurnal variation B or C remained the same, but that from other causes the general temperature fell so uniformly that it may be represented by a straight line, such as D, what would the resulting trace be like? This we can easily find graphically, by drawing the line D and taking it as the variable level on which to add B and C. To do this we have only to measure the values of B and C at different hours from the level of the line D at the same hour. For instance, the minima at 4 a.m. of E and F are the same distance from D that B and C are from A at the same hour; also at 8 a.m. and 8 p.m. the diurnal curves equally cross the line of general level, and the maxima at 2 p.m. are also at equal distances from that level. Now look at the resulting curves E and F. The latter, which is the stronger, is able, as it were, to reverse the general fall of the line D, and every one would recognize that there was a diurnal range. But then turn to curve E. Here the diurnal variation is so small that it can only deflect, but not reverse, the general line D, and thus we get the apparently impossible result that the thermometer may fall all day, and yet that there may be a very distinct diurnal maximum and minimum, which only modify the rate of fall of the general sweep of the curve. Now, this is exactly what happened on December 9, 1874, at Stonyhurst, as shown in our meteogram (Fig. 28). There the thermometer fell all day, but by joining mentally the points where the trace cuts the first and second midnights of that day, we see at once that there is a diurnal maximum about 3 p.m. in the general sweep of the curve. In this case the curve is so irregular that, though we can detect the fact of a diurnal maximum, we cannot measure the amount even approximately. When, however, the trace is more regular, it is obvious that from a curve like E we could infer D, and the maximum and minimum diurnal values of E by a method exactly similar to that which we employed to find A from B. In fact, given B, C, and D, we can draw E and F; while, given E and F, we can separate them into a general line D and diurnal variations B and C respectively.

If the line of general level is so irregular that it cannot be represented by a straight line for twenty-four consecutive hours, then we can no longer separate the general and diurnal changes, for we are unable to draw the general level, which is then a curved line.

We have taken our illustration from temperature-curves, as their diurnal changes are the simplest and most obvious; but the same general principles apply to every meteorological element.

In barograms the diurnal variations in Great Britain are so small compared to the general changes that the former can usually be neglected; but in the tropics the mercury often falls regularly 0·12 in. in six hours, while the general changes of an approaching distant hurricane reach half that amount. Then the discrimination between general changes and diurnal variation of pressure is of vital importance.

The question of sorting out various sources of barometric movement is so important in every branch of meteorology that we must give the subject in greater detail.

Barometric Rate.

The measure of the rapidity with which the mercury rises or falls is called the "barometric rate." This is usually expressed by saying how many hundredths of an inch, or tenths of a millimetre, the mercury changes in an hour.

In a variable climate like Great Britain, anything under 0·02 in. per hour is a low rate, while anything over 0·05 may be considered high. On only a few occasions in any year will 0·10 be exceeded, though as high as 0·20 has been recorded in exceptional cases. In the tropics diurnal variation alone may give a rate of nearly 0·02, and anything higher than this would be a warning of danger.

As many people have a very vague and inaccurate

idea of the relation of cyclone-motion and gradients to barometric rate, it may be expedient to give some further explanations. If we look at the diagram of a cyclone given in Fig. 29, it is obvious that the rate of fall of A's barometer will depend on three things—the velocity of the cyclone; the steepness of the gradients; and the position of the observer relative to the cyclone-centre. If we assume that a distance equal to A B would be traversed by the cyclone-centre in four hours, A's rate would be $\frac{0.1}{4} = 0.025$, as the distance between the isobars is 0·1 in.; and it is evident that this rate might be doubled by doubling either the velocity of the cyclone or the barometric difference between the isobars. Then, in the same cyclone, with the same velocity and gradients, the rate to E, who was in the line of the cyclone's path, would be much greater than that to A, who was more remote. This is manifest, because, as a cyclone is generally nearly round, the shortest line between two concentric isobars is along a radius of the circle. In this diagram the distance E D is about half A B, so that E's barometric rate would be double that of A.

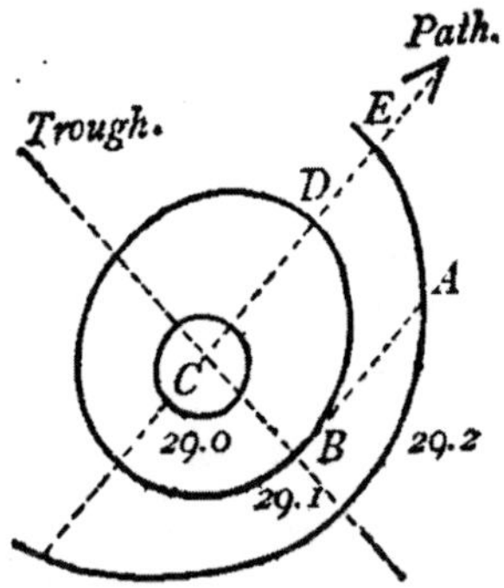

FIG. 29.—Barograms, barometric rate, and filling up of cyclones.

From these considerations we can see that a rapid fall of the barometer is dangerous, because, in a general way, it shows that the observer is nearly in a line with the path of the cyclone, that the gradients are steep, and that the disturbance is moving rapidly. The first gives an almost complete reversal of the wind, which is most dangerous to ships; the second, high wind; while the third increases the intensity of the weather in every way.

When we come to discuss squalls and thunderstorms we shall find that the barometer often jumps up 0·1 in. in a few minutes, just as the heavy rain begins. The cause of this is uncertain, but we must notice that the rise is of a totally distinct nature from that produced by the passage of a cyclone. We must not, therefore, be led into the error of treating all barometric changes as identical, and of comparing the barometric rate of a squall with that of a cyclone. A difficult case arises sometimes with the squall in the trough of a cyclone. As the wind goes round with gusts and heavy rain, the mercury turns upwards and rises with a sudden jump. Sometimes a slight fall then occurs, but directly afterwards the barometer rises quickly and steadily. It is almost impossible to say how much of the first jump is due to the squall, and how much to the general increase of pressure due to the passing on of the cyclone.

We have already explained that a barogram is a section, as it were, of a cyclone which is seen in plan on a synoptic chart. This, however, evidently only holds good on the supposition that the cyclone changes neither in depth nor shape during the time which elapses between the beginning and the end of the trace. Now, in practice a cyclone is perpetually changing both its depth and shape; and, consequently, it is often extremely difficult to see how the changes of pressure which are seen in two synoptic charts at even a short interval would have influenced the trace of the barometer at any one place. Two or more sets of general changes are going on simultaneously, and we have to work out the result of their combination. But the importance of investigating the question will be evident when we remark that on this depends all the apparently anomalous movements of the barometer. If the mercury always fell before rain, and rose when the weather began to mend, meteorology would be the simplest of sciences; but, unfortunately, we often see rain while the barometer is rising, or the sky begin to clear while pressure is still on the decrease. These exceptions have not been hitherto explained, but in this work we propose to do so.

Suppose that, as in Fig. 29, a cyclone, with centre at C, was moving along the crossed lane marked "path," at such a rate that it would traverse a distance equal to A B in four hours. Then—noting that the isobars at A and B are 0·1 inch of pressure apart; that the line A B is parallel to the path of the cyclone; and that B is on the line of the trough, where the barometer naturally begins to rise—if the cyclone moved onwards without any change in the depth of the centre, which is 29·0 in., the barometer at station A would fall 0·10 in. in these four hours, and then commence to mount.

But now, suppose that, while the centre moved onwards, the cyclone began to fill up at the rate of 0·16 in. in the four hours (and this is quite within practical limits), then the barometer at A would rise on balance, 0·06 in. in that time. In fact, it might be supposed to fall 0·10 in. from the approach of the cyclone's trough but to rise 0·16 in. from filling up, so that a gain of

0·06 in. would remain. Thus we explain the apparent anomaly of the barometer rising while a cyclone approaches; and here we see the enormous gain to knowledge which synoptic charts have effected. Formerly these barometric anomalies were inexplicable; now we can interpret them readily, for we know that rain and wind depend on the shape, and not the level, of the isobars. So that, though the cyclone is filling up and the barometer rising, the wind and weather at station A remain characteristic of the front of a cyclone. We shall refer to this subject again, and give a striking example in our chapter on Forecasting for Solitary Observers.

SURGE.

Any change of barometric level which is not due to the passage of some sort of depression or diurnal variation is called a "surge" of pressure. The word "wave" has often been applied to barometric changes, but in such an uncertain way that it seems best to coin a new word for a very definite and important phenomenon.

We have just explained the idea of a moving cyclone filling up, and of the resulting balance of a gain of pressure. It would have been just as easy for the cyclone to grow deeper in the same time, when we should have had the barometer falling in rear of the cyclone, with clearing weather. Sometimes filling up of a cyclone is tolerably local; other times surging is on an enormous scale. Nothing is more common in winter than to find a moderate-sized cyclone in mid-Atlantic one day, and that, though by next morning the shape of the isobars has hardly changed, the whole level of the cyclone and surroundings has perhaps decreased half an inch.

This hardly shows at first on a synoptic chart, for you see no change in the configuration of the lines; but, on looking at the figures attached to the isobars which denote the level, you see that what was 29·5 in. the first day is only 29·0 in. on the following morning. In like manner, a persistent anticyclone will often rise and fall one or two-tenths of an inch without any motion or material change of shape on the chart, while the barometer at any station will have appeared to rise or fall without any reason or apparent change of weather.

When we look at a series of these surges we find a decided tendency of the motion to travel from west to east, or from south-west to north-east. For instance, suppose that one day there was a deep depression with one or more cyclones in the United States, an anticyclone in mid-Atlantic, and a shallow set of depressions over Europe. We might find by next morning that the American cyclones were filling up, but that the Atlantic high pressure was lower in level, but unchanged in position, while the European system was practically unaltered. The third day might see that, with little change in America, the Atlantic anticyclone had regained its former level, while a great decrease of pressure had occurred over the whole of Europe.

This is the "barometric wave" of Birt and other writers, to which little importance is now attached, but which, the author believes, contains the germ of great development. The insuperable difficulty of tracing waves at present arises from the impossibility in most cases of separating the total barometer-change into its two components of surge and cyclone. Nothing is easier than to record the hour at which any barometer has touched its lowest point, but we cannot tell how much is due to the depression of a cyclone, or to the depression of a surge. It is manifest, from our last diagram, that we only observe when the balance is lowest.

A surge of itself has no characteristic weather, but the passage of a surge exercises a moderate influence on the characteristic weather of any isobaric shape, and a very powerful one on the formation of new systems.

Let us define the front of a surge as all the part where pressure is decreasing; the rear as all the part where pressure is increasing; the trough as the line of change from fall to rise; and the crest as the line of change from rise to fall. Then we find that the front of a moving surge, or the mere deepening of a cyclone, does not alter the typical character of the front and rear of the cyclone, but increases the general intensity; while the rising part of a surge decreases the intensity, and so improves the weather. The lowering of an anticyclone decreases the dryness and increases the tendency to form cloud, while gain of pressure has the opposite effect. But the most striking and by far the most important effect of surge is the influence on the development of new systems of disturbance. The tendency of all reduction of barometric level all over the world is to induce cyclonic systems, while that of gain of pressure is to dissipate existing cyclones.

We shall find abundant examples of this great principle in our illustrations of types of temperate weather.

There, as we have just mentioned, surge and cyclone are so mixed up together that we can only partially disentangle them; but in the tropics we find the same law under simpler conditions. For instance, in the South Indian Ocean, during the period of the north-west monsoon —from about December to March—there is a long furrow of low pressure about 10° south latitude, where the north-west monsoon meets the south-east trade. During the whole of that season this general depression goes through a series of small surges, gradually lowering, perhaps one-tenth of an inch for six or seven days, and then rising about the same amount in another week. Now, as a matter of observation, hurricanes almost invariably form during the downward period of the surge, and practical forecasters, like Meldrum at the Mauritius, are always specially on the look-out for signs of serious bad weather whenever there is the slightest symptom of a non-diurnal diminution of pressure. We believe that the same principle of watching surges might be used for forecasting with great advantage in temperate regions, in spite of the difficulties in the way of practical application, to which we have already alluded.

It is absolutely necessary, in dealing with such complicated problems as occur in meteorology, to have short simple terms to denote certain sets of phenomena. Here, and throughout this work, we shall talk of all changes of pressure due to the passage of an unchanging area of low pressure as cyclonic-changes, and all due to surging or re-arrangement of existing systems as surging changes; and we shall talk of surge overriding the cyclone, or cyclone overriding the surge, when, in our chapter on

Forecasting for Solitary Observers, we have to explain more fully many apparent anomalies in the behaviour of the barometer.

INTERPRETATION ON METEOGRAMS.

We must now examine still further the relation of charts to meteograms, and explain their respective values and interpretations.

Synoptic charts in practice can rarely be constructed more than three times a day, and it is obvious that, though general changes, such as the formation or motion of cyclones, can be shown on them with the greatest clearness, the nature of diurnal variations could not be properly discovered by their means. We shall illustrate this more fully in our chapter on Diurnal Weather, but here we must consider how to collate the variations due to the time of day with the general changes.

Almost all over the world the velocity of wind increases with the day and falls during the night, as we saw in our meteogram, Fig. 25, and this occurs both in cyclones and anticyclones. How can we collate this with the fact that, from charts constructed at the same hour on different days, the velocity is proportional to the different isobaric gradients? The answer obviously is, that if we suppose the gradient to remain unchanged for twenty-four hours, the mean velocity of the wind may be considered the speed due to that gradient, and that a diurnal variation of velocity for gradient is superimposed on this. For instance, suppose the wind to vary diurnally between ten and twenty miles per hour during any day, so that the mean velocity was fifteen miles, and the variation due to diurnal influences ten miles an hour; also that the gradient remained unchanged: then a synoptic chart constructed at the hour of minimum wind—about 4 a.m. —would give a velocity of ten miles an hour for the given gradient, while another constructed at the hour of maximum wind—say at 2 p.m.—would give a velocity of twenty miles an hour for the same gradient, and so on for every other hour.

The diurnal variation of direction introduces some other considerations. In the temperate regions of the northern hemisphere, the wind usually veers a little during the day and backs again at night, from whatever direction it may come. If, then, we consider the angle between the wind and the isobar, the above means that the angle is less by day than during the night. When we stand with our backs to the wind, we are generally at an angle of about 35° to the left of the isobar; so that if the wind veers, say, from south to south-south-west during the day, while the lie of the isobar remains the same, the angle between the wind and isobar would be diminished.

It has already been noticed that the wind in a cyclone is always incurved, while in an anticyclone it is outcurved. We therefore infer, from the fact of the mean diurnal veering of wind, that in cyclones the wind is a little less incurved, and in anticyclones a little less outcurved, by day than by night.

The following will illustrate the above principles. In Figs. 30, 31, and 32 we give a reduction of the United States daily charts for January 20, 1873, at

11 p m., together with that at 4.35 p.m. and 11 p.m. on January 21.

These charts may be taken as representing a freely moving cyclone, the intensity of which, as measured by the gradients, is pretty constant; but when we look at the wind-arrows it will be seen that, while in the two 11 p.m. charts there is only one station in the first where

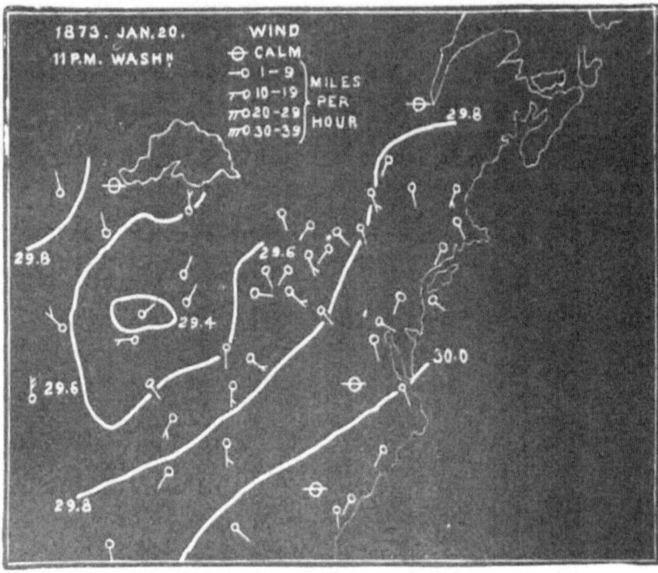

FIG. 30.—Diurnal variation of wind in a cyclone.

the wind exceeds twenty miles an hour, and none in the second, the 4 35 p.m. chart not only has three stations with that velocity, and one over thirty miles, but contains a far larger number of arrows indicating more than ten miles an hour, as shown by the feathers on the arrows. The original records show that the total miles of wind at all the seventy-five reporting stations in the first chart is 449 miles, with eight calms; in the second, 681 miles, with only five calm stations; while in the third chart, the wind has fallen to 420 miles, and the calms have increased to twelve, though the gradients remain pretty constant.

Next as regards the diurnal variation in the wind's direction. Though not very obvious, still, on the whole the arrows in the 4.35 p.m. chart will be found rather less incurved than in either of those at 11 p.m. relative

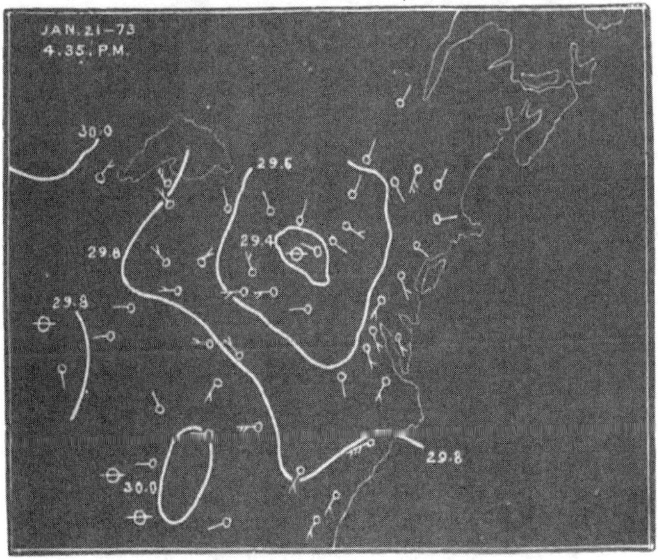

FIG. 31.—Diurnal variation of wind in a cyclone.

to the cyclone-centre; so that at every station the wind, from whatever direction it may blow, appears to veer a little with the sun during the day, and to back towards night, unless overridden by the greater changes due to the cyclone's motion. Similarly, if the charts had been constructed at the same hours for an anticyclone, the wind would have been found a little less outcurved at 4.35 p.m., and at every station the wind would also have veered a little during the day, and then backed towards evening.

The relation of the weather in isobars to the diurnal variation in the frequency of rain and cloud at different hours is rather more complicated. Suppose that all rain was cyclonic, and that the curve of mean diurnal frequency of rain showed a maximum at 2 p.m., and a

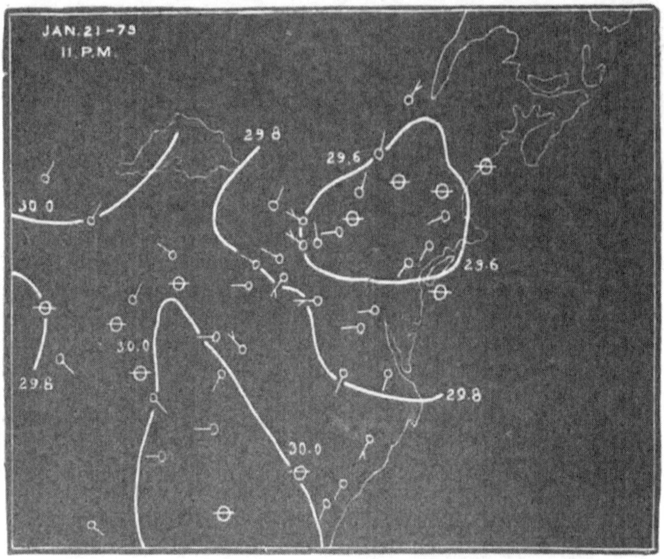

FIG. 32.—Diurnal variation of wind in a cyclone.

minimum at 4 a.m., what difference should we expect to see in a synoptic chart for any particular cyclone if it was constructed for those two hours? The inference undoubtedly is that the general position of rain and cloud relative to the lowest isobars would be unchanged, but that the rain and cloud would extend further from the centre at 2 p.m. than at 4 a.m. Thus, if we could conceive a stationary and unchanging cy lone, it would rain the whole twenty-four hours to an observer inside the diminished rain-area which the 4 a.m. chart would show, while it would begin to rain in the morning and cease towards evening to an observer situate anywhere between this and the outside of the 2 p.m. extension of the rain-area. As an illustration, we give in Figs. 33, 34, and 35, in a diagrammatic form, the position of cloud

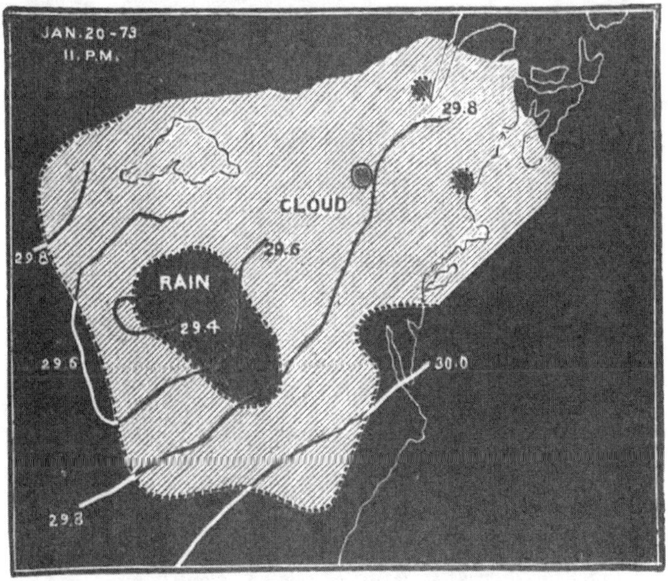

FIG. 33.—Diurnal variation of rain and cloud in a cyclone.

and rain in a typical cyclone of pretty constant shape and gradients in the United States on January 20, 1873, at 11 p.m., and on the 21st at 4.35 p.m., being the same cyclone whose winds have already been discussed.

From these it will be readily seen that the area of rain and cloud round the centre in the first 11 p.m. chart is considerably increased in the 4.35 p.m. chart, and again diminished in size in the second 11 p.m. chart. The portion of the outside bounding line of the cloud-area which is dotted shows where observation gave the end of cloud and appearance of blue sky. The portion where the single shading ends without a dotted line merely shows where observations ceased, and that the cloud extended to some unknown distance beyond these limits.

So far for the interpretation of the relations of the

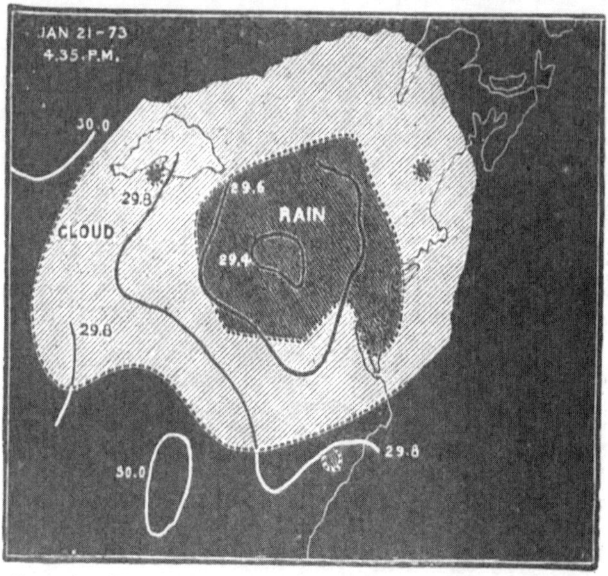

FIG. 34.—Diurnal variation of rain and cloud in a cyclone.

diurnal variations which we find in meteograms to the facts concerning the nature of weather which we derive from synoptic charts; but we must now consider some of the more minute phenomena of cyclones, etc., which can only be learnt from meteograms and from verbal descriptions of the sequence of weather on each day.

We have already explained the broad features of the sequence of blue sky, clouds, rain, back to blue sky again, in a cyclone; but it is obvious that, as the stations from which the materials for making synoptic charts are derived are rarely less than from eighty to a hundred miles apart, many of the details of a cyclone are lost. For instance, in practice, we rarely get more than one or two stations to report halo at the same time, and from that we could never deduce the shape of the halo-forming

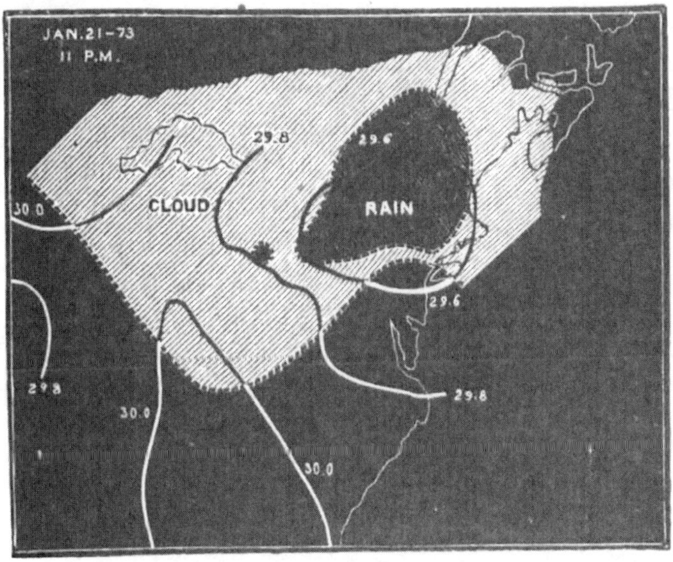

FIG. 35.—Diurnal variation of rain and cloud in a cyclone.

portion of a cyclone. But when we observe in a great number of cases that halo sky rarely lasts long, and is exclusively formed in front of the regular cloud-area, then we conclude that the ring of halo sky, such as we have marked in our diagram of cyclone-prognostics, is very narrow. The reason is, that as the cloud-area is propagated at the same rate as the cyclone, which we may take at twenty miles an hour, and that as a halo usually lasts, say, only half an hour, therefore the width of the halo-ring will not be much more than ten miles. Of course, if charts could be constructed for every hour of the day, and at stations only five or ten miles apart, there is nothing we learn from meteograms which we could not also derive from charts; but, as such observations are impracticable, it is of the utmost importance to know precisely how the continuous trace of instruments at any one station can be collated with the intermittent observations at widely scattered localities.

The most striking example of the value of meteograms in building up the nature of a cyclone is found in the phenomena of the trough. These are confined to a line only a mile or two wide, and it would be utterly impossible from charts alone ever to learn the significance of the turn of the barometer. We might look at fifty charts of different cyclones, and it might happen that the trough was not actually passing over any observing-station in any one of them.

But if, in a large number of cyclones, it is found that whenever the barometer turns to rise there is a squall, this, being independent of the time of day, must be referred to that part of the cyclone for its origin; and since this phenomenon occurs at all places over which the cyclone-trough passes, however distant from its centre, if a synoptic chart could be made with a large number of stations close together, a line of squalls would be seen under the trough of the cyclone, marking all the points at which the barometer turned to rise simultaneously.

This inference may be derived either by taking the history of the passage of a single cyclone, and observing that a squall was associated with the trough at every station, or else by observing at any one station that in every cyclone which passed, the trough and a squall came together. The latter deduction hangs on the assumption that in a great number of cyclones no two need be supposed to pass at the same distance from the station; so that, to a certain extent, a large number of different sections of a single cyclone, and a large number of single sections of different cyclones, give the same result.

The method of the meteorologist is, in fact, analogous to that of the microscopist, who builds up his picture of the organs of an animal by taking a series of their sections, across any portion of it.

There are many other deductions which can be made as to how the flexure of barograms indicate the nature of the gradients that are being propagated over any place; and as to how squalls and thunderstorms, and even single gusts, each leave their characteristic mark on a barographic trace, which can be read off any time afterwards by a practised observer. We have already explained how some of the fluctuations of a thermograph tell their own story about cold showers, or passing clouds, and many other deductions can be made from these traces. We could also point out how wind-traces reflect each fitful gust in their own appropriate manner; and also how minute details of the relation of wind-direction to cyclone and anticyclone centres, as well as minor diurnal variations both of force and direction, can be deduced more accurately from anemograms than from charts. The consideration of these is, however, unsuitable for an elementary work, and the object of this chapter so far will have been attained if we have conveyed to the reader a clear idea that observations at any one station give a section of the weather-changes which are shown in plan on successive synoptic charts; and that each self-recording instrument writes in its own language, and, as it were, in its own alphabet, the history of the weather for every day.

Descriptive, or Non-Instrumental, Records.

So far we have discussed the significance of instrumental records; but, however skilfully we may read those written traces, it is evident that there is still a great deal of weather about which they tell us nothing. No mechanical registration of pressure, temperature, or wind can ever make up for the want of a good verbal description of weather-sequence. No instrument can picture to us the various ways in which a blue sky can become overcast; whether the blue grows gradually pale and sickly, or whether great snaky-looking clouds seem irresistibly to embrace the whole heavens. Neither can it describe the delicate distinctions which our senses enable us to perceive in the way the wind blows. Our eyes tell us at a glance that a south-west wind raises a long sea, while a nor'-wester rakes the surface of the ocean into lines of foam; and that the fitful gusts of an impending shower drive little eddies along the dusty road.

In like manner, no short cloud-symbols, such as detached cloud, overcast, misty, or even the more detailed words—cirrus, cumulus, etc., can ever give more than a lifeless picture of the sky as we know it.

The old myth-makers excelled in their descriptions of weather. In their own peculiar figurative language we see reflected a vivid picture of cloud and thunderstorm which we can scarcely match in the more sober verbiage of modern times. The Greek poets knew the difference between the beneficent diurnal winds which sprang up at dawn and the dangerous blasts of an approaching thunderstorm; and never mistook the wind which sighed among the pinetops for the north-westerly squalls which tumbled the trees over the cliffs.

All that instrumental traces could tell of this would be deduced from seeing if the velocity-trace had some connection with the time of day, or if it was fitful, and that the direction-trace was also unsteady; or whether some directions, such as the north-west, were associated with higher velocities than others.

On the other hand, instruments not only give precision to the general impressions derived from the senses, which alone a savage can receive; but also enable us to discover some changes which our perceptions alone could never detect. For instance, by measuring heat-curves we can calculate the ordinary amount of daily range, and compare the value in London with that in Berlin, or New York; and we can also draw deductions from certain bends in the temperature-curve which would never have entered into the head of semi-civilized man. In like manner, there is in England a small increase of the wind-velocity about 1 a.m. which has some scientific interest, but which certainly would not have been discovered without instrumental appliances.

But, in addition, the invention of the barometer has

given us another sense—that is to say, the appreciation of the varying weight of the atmosphere, which was denied to our ancestors; and this book is the answer to the question how much weather-knowledge can be derived from observation of that instrument.

It will be found a distinguishing feature of this work that we have endeavoured to describe the weather in different shapes of isobars, so far as possible, in the language of popular prognostics. This language, while it contains many survivals of mythic speech, is still in current use, and gives a much more accurate picture of weather than more formal language. It is far more life-like to talk of a cyclone-front as dirty and muggy than to report sky overcast, humidity ninety-eight per cent.; or to say that the sun "draws water" in straight isobars rather than c. 9 stratus (sky nine-tenths overcast, stratus-cloud). We use, in fact, the phraseology of popular weather lore to translate, as it were, the indications of instrumental readings into the language of common life.

At the same time, we have already examined most carefully the minuter fluctuations of some instrumental traces, and in various chapters we shall investigate the precise significance of the results of various arithmetical calculations which can be made from the numerical values derived from thermograms, etc.

The problems which the meteorologist has to solve are so complex and varied that he cannot afford to dispense with any possible assistance from whatever quarter; and our endeavour has been to convey to the reader the results of every line of investigation, and to collate the old and new meteorology into one compact science.

CHAPTER VI.

WIND AND CALM.

In the preceding chapters we have only stated that in most cases the force or velocity of the wind is roughly proportional to the closeness of the isobars; but we shall now go into the details of the subject, and give the actual numbers which connect wind and gradients. We shall then point out various sources of variation which prevent us from laying down any law of wind with mathematical accuracy, and carry out the same idea with reference to the relation of the angle between the direction of the wind and the lie of the isobars. After that we shall extend these and other general principles of wind to the southern hemisphere, and conclude with a few general reflections on the subject.

Gradients.

The relative closeness of any two isobars is not measured by the number of miles between them, but by the steepness of the barometric slope which they indicate. For instance, suppose that two isobars differ by 0·2 in. (5 mm.) of barometric level—say 29·7 and 29·9 in.

(755 and 760 mm.)—we do not measure their relative proximity by saying that they are thirty or ninety miles apart, but we think of the barometric slope with a rise of two-tenths of an inch (5 mm.) in either thirty or ninety miles. Then, to reduce this to a common standard, we take a uniform distance—in England fifteen nautical miles, or seventeen statute miles—and calculate how many hundredths of an inch the barometer would rise in those fifteen miles; that is to say, we treat the barometric slope like the slope of a hill, which is universally estimated by saying that the latter rises so many feet in a mile.

The slope between two isobars is called the barometric gradient, and, of course, it is measured square or at right angles to the isobars, just in the same way that we measure the slope of a hill between two contour lines.

For instance, suppose that in Fig. 36 the line A B, drawn square to the isobars, is thirty nautical miles long, and that the isobars denote differences of two-tenths or twenty-hundredths of an inch; then the rise in fifteen nautical miles would be ten-hundredths of an inch; and we should say that there was a gradient of ten between the two stations A and B. If the distance between the same two isobars at C and D was ninety miles, the gradient over an observer at E would only be $0{\cdot}2 \times 100 \times \frac{15}{90} = 3{\cdot}3$; and this last number would be the required gradient.

In practice we too often come across the error of taking the difference of pressure at two places, F and G, and calculating the gradient from the distance in miles between them. This always gives a smaller gradient than the real one, for the line of a gradient is always the shortest line which can be drawn between two isobars. In this case G H, not G F, is the proper line to measure for the gradient at G. The best method for measuring gradients on a synoptic chart is to take the station E, whose gradient

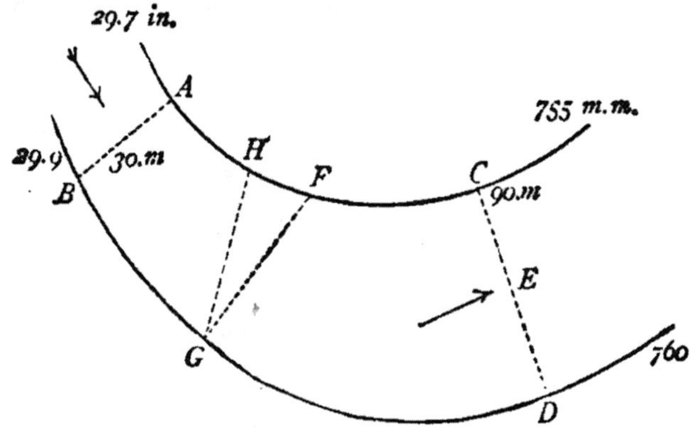

FIG. 36.—Barometric gradients.

is required, and then to draw the line C D through E on the map, square to the isobars, and measure its length on the scale of the chart. This is much quicker and more accurate than trying to find two places which are nearly square to the isobars. Gradients as thus measured are rarely higher than four or five in Great Britain, though much higher values than these are occasionally recorded in very severe storms. In a general way, it may be said that gradients are slight or moderate when below one, and steep when above two. In Europe gradients are measured by the number of millimetres of barometric difference in one degree of a great circle of the earth; that is, in sixty geographical miles. For any slope the numbers are nearly identical with those in England, for one millimetre very nearly equals 0·04 in., and fifteen nautical miles are exactly equal to one degree.

RELATION OF VELOCITY TO GRADIENT.

But it may be interesting to see what the velocity of the wind actually is for any given gradient.

The following are the numbers obtained by Messrs. Whipple and Baker at Kew, near London, both in English and metrical equivalents:—

Gradients per fifteen nautical miles.	Wind-velocity in miles per hour.	Gradient in millimetres per degree of latitude.	Velocity in metres per second.
0·2	5·0	·203	2·23
0·5	7·0	·508	3·13
0·7	7·5	·711	3·35
1·0	9·2	1·016	4·11
1·2	11·6	1·219	5·19
1·5	12·6	1·524	5·63
1·7	15·0	1·727	6·70
2·0	16·5	2·032	7·38
2·2	19·1	2·235	8·54
2·5	22·0	2·540	9·83
2·7	22·2	2·743	9·92
3·0	25·2	3·048	11·40

Loomis has arrived at the following values in the United States:—

Gradient millimetres per degree.	Velocity in metres per second.	Gradient millimetres per degree.	Velocity in metres per second.
2·09	7·20	2·80	10·64
2·31	8·18	2·90	11·09
2·48	9·03	3·08	11·44
2·61	9·66	3·36	11·80
2·72	10·28	3·73	12·20

These agree very fairly well with the British observations.

In the Atlantic, Professor Loomis finds that for the same gradient the velocity of the wind is forty per cent. greater than in the United States. This is doubtless due to the influence of a certain number of sheltered stations among the land-observatories. Nearly every place feels some winds more than others, and will therefore sometimes report comparatively little wind for a considerable gradient. Thus, when the results are averaged, the mean values will be lower than if every observer was equally exposed to all winds, as in the open sea.

Wind is much stronger for the same gradients in the tropics than in higher latitudes. In the Indian Ocean, especially, the north-east and north-west monsoons blow steady with a gradient that would give variable winds in temperate regions; and the violence of the south-west monsoon is out of all keeping with the steepness of gradient according to European experience.

Variations in Velocity and Gradient.

There are various sources of variation from these general laws of the relation of velocity to gradient, some of which only can be explained in an elementary work like the present. In Great Britain it is found that, for any given (moderate) gradient, winds from north and east points are stronger than those from south and west points. For instance, Ley has found at Kew the following differences:—

Gradient per fifteen nautical miles.	Velocity in miles for winds from S.S.E. by S. to N.W.	Velocity in miles for winds from N.N.W. by N. to S.E.
·006	4·14	6·89
·009	6·41	8·63
·012	8·37	10·93
·015	11·21	14·27
·016	13·56	16·98

The reason why the mean of these does not agree with the mean velocity for the same gradients as deduced by Mr. Whipple for the same station is readily explained. The latter takes all winds from all directions for the same gradient, and averages them up together. Mr. Ley separates the two principal directions; but, as one direction—the south-west—is much more frequent than the north-west, his numbers would have to be weighted proportional to the frequency of these two directions to give the same numbers as Mr. Whipple.

Local variation of wind is too obvious to need much comment. The only thing we have to consider here is how it affects forecasting. If every station was equally exposed to every quarter, it might be possible to issue forecasts in which the amount of wind recorded by any instrument might be approximately indicated; but when we have to deal with a gale which begins in the south-east and works round to north-west, it is manifest that we can only state the probable amount of wind in general terms, as no place is equally influenced by wind from these two quarters.

The relation of diurnal variation of wind-velocity to

gradient has already been discussed in our chapter on Meteograms; and other diurnal phases of wind, such as land and sea breezes or valley winds, will be most conveniently taken in our chapter on Diurnal Variations of Weather. But, in temperate regions, by far the most important elements of disturbance in the simple relation of wind to gradient are squalls and thunderstorms. In both of these the barometer usually rises suddenly, sometimes as much as one-tenth of an inch, from causes which are at present obscure. And in both we find angry, violent gusts, which bear no relation whatever to isobaric gradients. Many of the discordant observations on this subject are doubtless due to want of care in distinguishing one kind of wind from another.

Loomis has called attention to the total want of accordance between wind and gradient which he has found during the "northers" of New Mexico, and the author has found that in the "nortes" of Panama the wind also is quite disproportionate to the gradient. Finley has also discovered in the United States what are called "straight-line gales," or long streaks of wind, two or three miles across, blowing at the rate of sixty to eighty miles an hour, and extending over eighty or one hundred miles in length. These appear on the side of a cyclonic depression, at some distance from the centre, and are not associated with any deflection of the isobars. Other winds that are not directly associated with isobaric gradients have been noticed in other parts of the world, and we are therefore led to the conclusion that, though the great wind-circulation of the atmosphere is related to isobars, still there are some winds that are impelled by other causes than those which develop isobars; and for the sake of classification we will call them generically "non-isobaric winds." They are most probably connected with what we have before alluded to as non-isobaric rains.

We cannot say what is the origin of the wind in thunderstorms and non-isobaric winds, but it is certain that the cause is quite different from that in cyclones. We must therefore take care, in talking about wind, not to mix up two kinds which really have little in common. From all this we see the very fallacious results which come of trusting blindly to instruments, and also that any statistical values which are derived from mixing up various sorts of wind can only give rise to discordant deductions.

We may also remark that merely saying that a storm blew with such a force or velocity tells us very little either of the true character of the wind or of the amount of destruction which the gale might cause. An instrumental record of forty miles of wind in an hour may be made up either of a steady weight of wind, or of violent gusts alternating with quieter intervals. The damage done in the latter case would many times be greater than in the former.

Then there are many minute differences in the way of blowing which instruments cannot even detect. We all know that most chimneys smoke more with an east wind than with a west one. We have also just shown that the velocities of these two winds is not the same for the same gradients.

It has been suggested, with a great deal of probability,

that the difference may be due to the wind not blowing horizontally, and that east winds are perhaps directed a little downwards. Another very striking phase of wind is the difference between the kind of sea raised by the south-west gale in front of a cyclone and the north-wester in rear. The first raises a high sea with only a moderate amount of white water, while the latter rakes the surface of the ocean into long streaks of foam. There are other reasons for believing that in front of a cyclone the wind is rising, while in rear the air-currents have a slant downwards. If so, the cold, dry clearness of north-westers is readily explained. The whirls of dust that precede some kinds of rain are also familiar instances of the specific character which belongs to different winds.

Relation of Direction to Gradient.

We will now consider the details of the relation of the direction of the wind to gradients and to the lie or trend of the isobar conjointly.

When we talk of gradient only, we get no indication of the direction of the wind, for the barometric slope may face in any direction or have any aspect. Following the analogy of barometric gradients to hill-slopes, we will call the direction towards which gradients slope the aspect of the gradient, so as to keep the word direction for wind. For instance, if isobars run east and west they may slope either north or south, or we might say that the aspect of the gradients was either towards the north or the south, just as we should talk of a hill; or, to take the analogy of geological terms, we might say that the strike of the isobars was east and west, but that the gradients dipped either north or south.

But by combining the idea of gradient with that of aspect, and both with the Buy Ballot's law, we see at once that if the isobars run or strike east and west, the general direction of the wind will be westerly if the aspect of the gradients is towards the north, but easterly if the aspect is to the south. We therefore say that in the former case we have gradients of such a value for westerly winds, and in the latter gradients of such an amount for easterly winds. This holds for every direction. In Fig. 36 we have, as before explained, a gradient of ten between A and B for velocity, and now we can say that the gradient is also for north-westerly winds; at E there is a gradient of 3·3 for south-westerly winds. By this simple method of expression, whenever we see a synoptic chart, we can calculate at once both the probable direction and force of the wind.

Inclination of Wind to Isobars.

Buy Ballot's law does very well for the general sweep of the wind, but the subject is capable of much greater refinement. The acute angle between the direction of the wind and the lie of the isobar is called the inclination of the wind to that isobar. Taking all kinds of winds and all kinds of isobars, Whipple has found that the inclination amounts to 52° at Kew; while Loomis has deduced an angle of 42° in the United States.

But, by taking the inclination of the wind in different shapes of isobars and different portions of each shape,

Ley, Loomis, Hildebrandson, and others, have arrived at a series of remarkable generalizations as to the general circulation of the atmosphere. They find that the wind is much more inclined and incurved in the right front of a cyclone than in any other portion; and that in the rear the inclination is very small, if not occasionally reversed —that is to say, a little outcurved.

We have examined the details of these cyclone surface-winds, as well as of those in an anticyclone in our chapter on Clouds. There we treated each shape separately, but we can connect both in a very striking manner if we call attention to some general values obtained by Loomis from observations over the Atlantic Ocean. Taking an ideal cyclone, with an adjacent anticyclone, he finds that, starting from the anticyclone, the inclination of the wind to the isobar begins at about 52°, and then gradually decreases to 25° near the centre of a cyclone. Of course this is a generalized case, for we have shown that the inclination is not the same on different sides of a cyclone. The great thing to remember is that in every shape of isobars each part has a wind velocity and direction of its own relative to the gradients.

The only other material source of variation is diurnal. We have already sufficiently explained, in our chapter on the Meteograms, that, whatever the inclination due to any part of any shape of isobars may be, the diurnal variation imposes a modification on that, but does not alter the direction due to general causes. Land and sea breezes we shall discuss in our chapter on Diurnal Variations of Weather.

Calms.

We have already stated that calms are the product of no barometric gradient. The most persistent calms are found in the "doldrums," or the col of low pressure near the equator between the north-east and south-east trade winds all over the world.

In temperate regions the most persistent calms are near the centres of stationary anticyclones; but more short-lived calms are found in the centres of cyclones, along the crest of wedges, and in cols.

We do not think it necessary to give any special examples of either gales or calms, for they are abundantly illustrated by numerous charts in the course of the work; we need only call attention to Figs. 65 and 66 of south-westerly gales in Great Britain, to Figs. 77 and 78 of easterly gales, and to Figs. 22 and 24 of calms.

Winds in the Southern Hemisphere.

So far we have confined our attention to winds in the northern hemisphere only; now, however, that we thoroughly understand the nature of wind in that hemisphere, we can easily follow the modifications which occur south of the equator.

The great general principles—that every shape of isobars has a distinctive wind; that cyclones incurve, while anticyclones outcurve; that the velocity is mainly determined by the gradient, and also the relation of diurnal to general winds—are the same in both hemispheres. What does differ is that portion of Buy Ballot's law which gives the position of the nearest low pressure to an observer who turns his back to the wind. For the southern hemisphere the law is as follows:—stand with your back to the wind, and pressure will be lower on your right hand than on your left. This is exactly the converse of what holds north of the equator.

As a necessary consequence of this, the surface-wind will rotate round a cyclone or anticyclone in the opposite manner to what it does in the northern hemisphere. That is to say, a cyclone rotates in the direction of the motion of the watch-hands, but is incurved; while the anticyclone turns against the watch-hands, but is still outcurved, as in Europe. The general circulation of the upper currents is exactly analogous to that of the northern hemisphere, being nearly parallel to the isobars at the level of the lower cumulus, and more or less outwards at the higher cirrus level in cyclones, and inwards in anticyclones.

For this reason, the vertical succession of the upper currents is contrary to that of Europe. There the upper currents always come successively more and more from the left as you stand with your back to the wind; whereas they will come more and more from the right in the southern hemisphere. For instance, if the surface-wind was south, in Europe the lowest clouds would be south-south-west, the next layer south-west, and the highest cirrus perhaps from west. Whereas in Australia, with the same surface-wind from south, the successive upper currents would come from south-south-east, south-east, and east respectively.

The diurnal variation of wind-velocity will be fully discussed in our chapter on that subject.

The sequence of wind as a cyclone drifts past an Australian station would be different from that of Europe. Anywhere south of the line, the wind goes round by the north if the centre passes south, and round by south if the centre passes north, of the observer, which is exactly the converse of what happens in Europe. Mr. Ringwood has pointed out that we can express both cases by one general law, if we say that in both hemispheres the wind goes round by the polar side when the centre passes on the equatorial side of the station, and by the equatorial side when the centre passes on the polar side.

This can all be better illustrated by a few actual examples than by generalized diagrams, the more so as the figures can then be made to show some other interesting phenomena.

In Fig. 37 we give an example of a violent cyclone which blew in the Indian Ocean on February 13, 1861, between 10° and 20° south latitude, and about 80° east of Greenwich, as deduced by Mr. Meldrum, of the Mauritius. The general nature of the rotation of the wind with the direction of the watch-hands will be very obvious, but we should note that the incurvature in most places is very considerable, especially to the west of the centre. To the south of the centre the wind was south-east; to the west, south-west; to the north, north-west; and from north-north-west to north-east on the eastern edge of the cyclone. The four feathered arrows denote a wind of hurricane force, and there is nothing in the steepness of the gradients to suggest such high velocities.

The path of the cyclone is marked by dated crosses, and we see that the motion, as usual in these latitudes, was very slow. From the 12th to the 13th the travel was a very little towards the east, at a rate of hardly three miles an hour; after that, the path was irregularly towards the south, at a rather higher speed.

We have introduced this example from the South

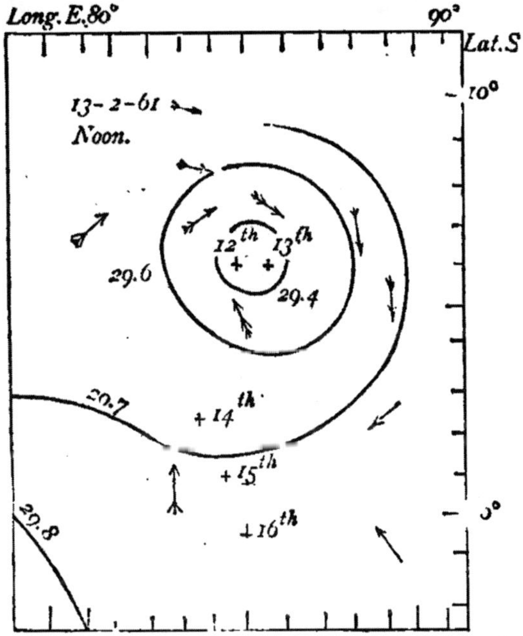

FIG. 37—Tropical hurricane (south of the equator).

Indian Ocean partly to show that in all principal characteristics a tropical does not differ from an extra-tropical cyclone; but we shall understand the antithesis of the wind-sequence far better from an Australian example, because the conditions of weather in that country are more similar to those of Europe or the United States than those of the lower latitudes.

In Figs. 38 and 39 we give the wind and isobars for Australia on November 20 and 21, 1884. For these we are indebted to Mr. Ellery, of the Government Observatory at Melbourne. We find in the first chart (Fig. 38) that the highest pressure is over Queensland, and that a moderate-sized cyclone covers the Australian Bight. The wind rotates round this in the usual manner of the

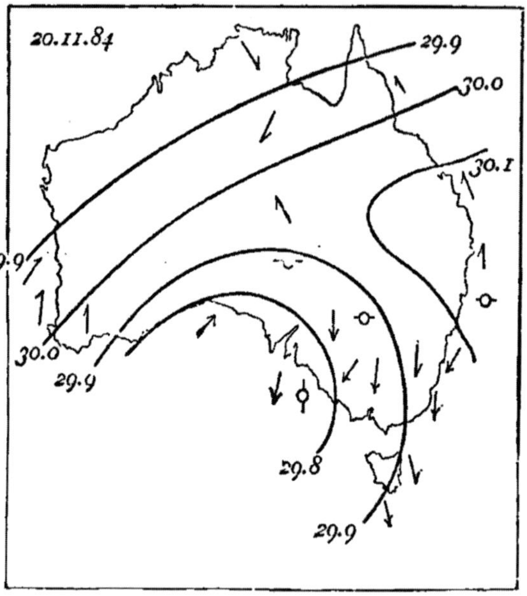

FIG. 38.—Cyclone (Australia).

hemisphere, being north and north-east in front, and south and south-west in rear. Land and sea breezes deflect the winds along the coast-line, and in the interior of the island variable winds are reported, owing to the slight gradients which are there present. By next morning (Fig. 39) only the fragment of the cyclone appears to the south of Tasmania; portions of two anticyclones lie over the north-east and south-west corners of Australia; and a sort of ill-defined V-depression runs up towards the col which lies between the two anticyclones. The rotation of the wind round this V is from north-east and north in front, to south and south-west in rear, so that the wind-sequence at Melbourne for the day was from north-east

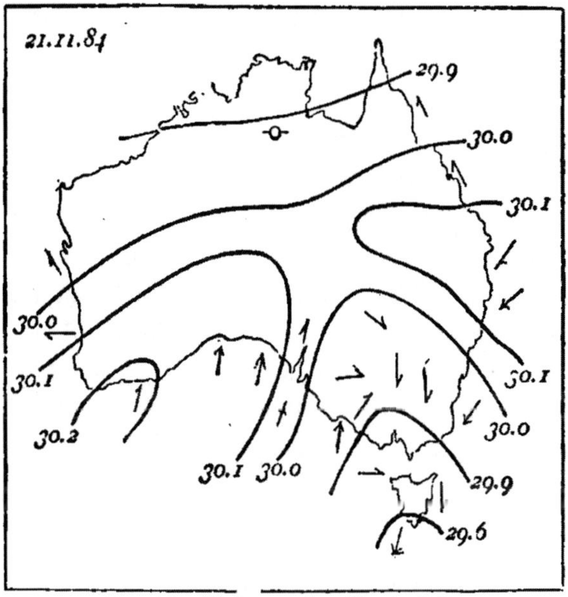

FIG. 39.—V-depression (Australia).

by north to north-west and south-west. In London the passage of the same isobars would have been associated with a shift of wind from south-east by south to south-west and north-west.

GENERAL REMARKS.

We may conclude this chapter with a few general remarks on the subject of wind.

In the first place, let us notice how little influence the rate of a cyclone's motion has on the velocity of the wind. All that we know for certain about the influence of the motion of a cyclone is that a high rate increases the general intensity of the wind and weather everywhere, but that it does not prevent the centre from being calm, or the wind from being light on any side where the gradients are slight.

In squalls the independence of the velocity of wind to that of the squall as a whole is still more curious. The latter may be travelling, perhaps, only twenty miles an hour, but the first blast may come at the rate of sixty miles an hour. This fact we must consider as due to an impulse being propagated which induces wind of such a velocity, but not as due to wind, or a gust, moving solidly over the earth's surface. Such an impulse is found in the trough of a cyclone or V-depression.

We have not thought it necessary to give the general principles of the dependence of wind-circulation on the earth's rotation, as that may be found in any text-book of physical science. The modification of Halley's old theory of north-east and south-west winds, which has been proposed by Professor Ferrel, has been universally adopted all over Europe and the United States. The theory is hardly known in England, and is too mathematical for this work. No doubt the earth's rotation is the real cause of the general direction of circulation in cyclones of either hemisphere, but what we cannot explain is the inclination of wind to the isobars. Theoretically, any small difference of temperature should set up a wind from the cold to the hot area; but we have seen already, and shall see still more in our next chapter on Heat and Cold, that differences of temperature even over large areas have wonderfully little influence on wind. The most that local differences of heat and cold do is to set up local breezes, such as land and sea, or valley winds. Then, theoretically, this cold wind should flow nearly straight toward the hot area, only a little deflected to the right or left, according to circumstances, by the earth's rotation. In like manner, any difference of pressure, from the high to the low barometer, however caused, should draw wind nearly straight. But, in our chapter on Diurnal Weather, we shall find some land and sea breezes which blow nearly parallel to the coast-line.

On the other hand, if we look at a cyclone purely as a circulating mass of air, the wind should be parallel to the isobar, perhaps even a little outcurved from centrifugal force. Now, in practice the wind is always incurved, and the depression of a cyclone is certainly not caused by centrifugal force. The fiercest wind which ever blew would only depress the barometer a few hundredths of an inch, instead of which we find depressions of two inches and more with no wind over fifty miles an hour. This, of course, is on the supposition that whirling air acts like a fluid.

The idea has been suggested that the friction of the wind on the earth's surface is the cause of the incurvature, and that without friction the wind would be parallel to the isobars, as we find it at the level of the lowest cloud-layers. It is extremely probable that this is at least partially true, for several experiments can be devised with whirling water, in which friction of small particles on the bottom does cause them to be collected in the centre, instead of being thrown out to the edges of the vessel.

Relation of Force to Velocity.

Lastly, we may say a few words about the relation of force to velocity. The velocity of wind is a real quantity, which is perhaps capable of measurement in the abstract, though we are at present far from being able to gauge it accurately. But it is quite certain that there is no such thing as an absolute force which corresponds to a given velocity. According to the theory of stream-lines, when even an inelastic fluid meets an obstacle, if the angles of the obstruction do not break the continuity of the fluid so as to form eddies or vortices, the same amount of pressure which is imposed on the body by the first deflection of the fluid is given back again as the stream-lines of the fluid close up behind the obstruction. For instance, if a ship is lying at anchor in a current, the same amount of strain which the current causes on her cable when forced asunder by the bows, is given back when the current closes in behind her; so that the total pressure which she experiences is only that due to the friction of the water on her skin. This is, of course, on the supposition that her lines are so easy that they do not break the stream-lines so as to form little eddies or vortices.

Now, the same thing holds with wind. If we put up two square plates of different sizes, face to the wind, the pressure on each is not proportional to the area, while in light breezes neither will record anything. The reason is that, in light wind, a thin mobile fluid like air can glide round even the sharp angles of a square without forming eddies, and as there is no vacuum formed behind the plate, there is no pressure recorded. In higher winds, the stream-lines are broken, and every shape and every sized plate of the same shape form a different series of eddies round the rim of the obstacle. Then the amount of rarification behind the various plates is neither identical nor proportional, and therefore every shape and size of anemometer indicates discordantly at every different velocity.

From all this it follows that, though we might say that the pressure on a board one foot square was twenty pounds, and might compare this force with that on another board of the same size and mounting, we should not be justified in saying that the force of the wind was twenty pounds per square foot in the abstract, because a board ten feet square, even if of the same shape, would have given a different number.

CHAPTER VII.

HEAT AND COLD.

In this chapter we purpose to go a little more into the details of the manner in which changes of temperature are produced. What are the causes of burning heats and hard frosts; why is the same day of the month hot in one year and cold in another; why at the same season do hot and cold days follow one another without any apparent sequence; and why is England sometimes warmer than France, though the latter is nearer the equator?

All these questions we propose to answer, and to point out how easily they can be explained by means of synoptic charts. The difficulties of getting rid of annual and diurnal variations have tempted many meteorologists still to adhere to the old method of averages, which can only lead to unsatisfactory, if not to delusive, results.

Diurnal Isotherms.

The question we have to solve is this. We know that the sun is the principal source of all heat, and, if nothing disturbed his rays, there would be a regular diminution of temperature from the equator to the pole, which we shall call a thermal slope. Every day, as the earth turned under the sun, a well-defined wave of variation would be imposed on this slope. The lines which mark out this deflected slope we will call diurnal isotherms. We have first to determine what the shape of the lines would be at any moment over the globe, and then how these diurnal isotherms would modify the appearance on a synoptic chart of any local developments of heat or cold. If we see on a chart that at eight o'clock in the morning there is a curious patch of heat in front of a cyclone, how are we to discover how much is due to cyclonic causes and how much to diurnal variations? In fact, how can we prove that the heat is due to the cyclone, and not to the time of day?

First, to form a conception of the diurnal distribution of temperature over the world at any moment. The author has shown that the general shape of the isotherms in any latitude would be like the lines in Fig. 40, if there was a uniform thermal slope from the equator to the pole, and no disturbing influences, such as unequal distribution of land and sea.

The diagram gives the ideal shape of the isotherms at any moment. Noon is placed in the middle of the diagram in longitude 180°, and the lines represent the diurnal variation in the latitude of each isotherm. The scale on the right of the diagram is degrees of latitude; that on the left, degrees of temperature. For an ideal diagram we have supposed that the general thermal slope from the equator is 1° of temperature for 1° of latitude. The principle on which the diagram is formed is as follows. Suppose that in latitude 20° the temperature is 60° Fahr. at midnight, and that by 6 a.m. the temperature has fallen to 59°; then we should have to go one degree of latitude further south at that hour, if we want to follow the position of the isotherm of 60°. If, in our undisturbed world, we could walk round the earth in any latitude in twenty-four hours, the line marked 60°

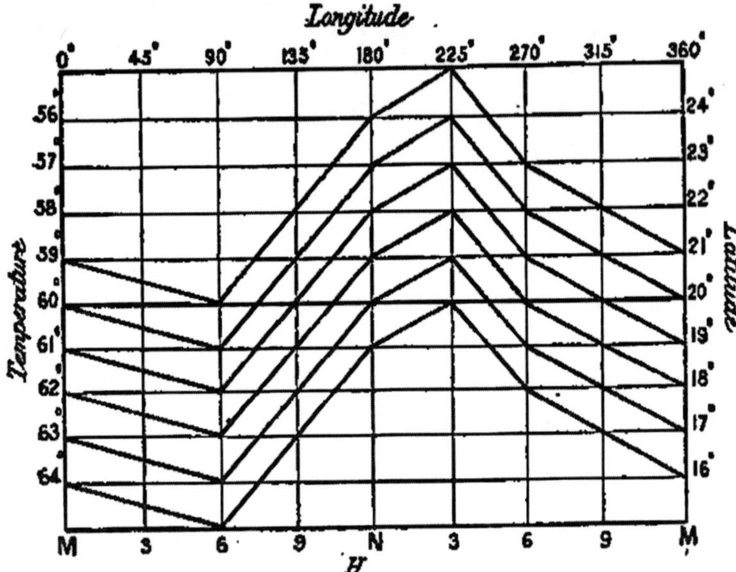

FIG. 40.—Diagram illustrating the shape of diurnal isotherms.

on the chart represents what our journey would be if we wanted to keep under a uniform temperature of 60° for the whole day. Starting at midnight on the left of the diagram, we should have to go sixty geographical miles south and 90° east of longitude by six o'clock in the morning. Between then and 3 p.m. we should have to make 300 miles of northing and 155° of easting, if we still wished to keep our thermometer at 60°; and from then till the second midnight we should have to make 240 miles of southing and 135° of easting, to follow the isotherm of 60.° Observe that the easting has to be expressed in degrees of longitude, for the number of miles in a degree varies with the latitude. The diagram is also based on the supposition that there is a pretty uniform isothermal slope from the equator to the pole, and that the diurnal range of temperature does not vary much within 5° or 10° of latitude.

Then, if there were no irregularities caused by cyclones, or the unequal heating of land or water, the diurnal thermogram in every place would be very similar in shape to the trace of any of the isotherms as plotted on a chart if we turn longitude into time, and latitude into degrees of heat on a suitable scale. In fact, we may conceive the curves shown in the diagram to sweep round the world with the earth's rotation, and suppose that the rise or fall of temperature at any station was caused by the passage of this shape of isotherms, just as the motion of the barometer is the product of the propagation of different shapes of isobars over any place. For instance, in the diagram (Fig. 40) the strong horizontal line shows the position of the section across the diurnal isotherms which is propagated over any station in latitude 20° north. Starting from the first midnight on the left of the diagram, the thermometer would mark 60°. By 6 a.m. the mercury would have fallen to 59°, as that isotherm descends to latitude 20° at that hour. Between 6 a.m. and 3 p.m. five isotherms are propagated over the station, so that the instrument would register 64° at the latter hour. Then, as lower isotherms begin to pass over the observer, the temperature would fall at the rate shown in the figure, till 60° was reached again by the second midnight.

How Diurnal modify General Isotherms.

Now, assuming this typical distribution of heat, we can readily see how the diurnal range of temperature modifies any isotherms which we find on a synoptic chart.

But, first, let us define the aspect of the thermal slope on the map of the world as the direction in which the gradients look, if we suppose the isotherms really to represent relative heights. For instance, in all curves the aspect of the slope in the morning after six o'clock is towards the north-west, while in the afternoon it is towards the north-east.

Now, suppose that at any hour we find a certain shape of isotherms on a synoptic chart: these lines represent the diurnal isotherms as modified by local radiation, etc.; or we may say that we have on the map temperature-distribution due to radiation or cyclonic causes lying on a diurnal thermal slope. Then, so long as the direction or aspect of the diurnal slope does not vary, the shapes of the isotherms will not alter; only the numbers which are attached to them will change. That is to say, the propagation of a uniform slope alters the level, but not the shape, of the isotherms.

For instance, let the square A B *b a*, in Fig. 41, represent an area of, say, 30° latitude by 30° (two hours) of longitude, anywhere on the surface of the earth, and let the slanting dotted lines mark a very exaggerated afternoon thermal slope, from 45° to 50° at the rate of 1½° per hour.

Also, suppose that the wind, etc., of a cyclone within the square had very much contorted the isotherms; the resulting shape would be compounded of the cyclonic disturbance lying on the simple diurnal slope shown on the diagram.

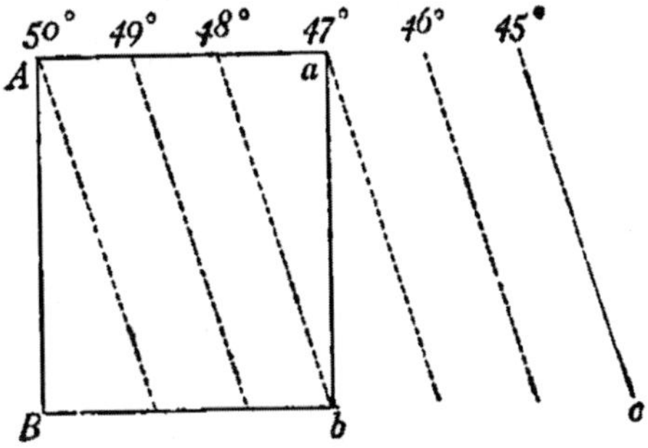

FIG. 41.—Thermal slope, and shape of isotherms.

Now, if, while the cyclone stood still, and the diurnal thermal slope was propagated over the square for two hours, then the only effect would be to leave the shape of the isotherms absolutely unchanged, but to make each line mark 3° lower. The isotherm of *a* would have arrived at A, *b* at B, *c* at *b*, and so on. That is to say, the lines marked 50°, 49°, 48° would be in the same places, but would be numbered 47°, 46°, 45° respectively, though the shape of the contortions would be the same.

If, on the contrary, the direction of the diurnal slope had changed during these two hours, from north-east to north-west—that is, from an afternoon to a morning aspect—then the shape of the contorted isotherms would have been much modified.

This conception of the propagation of a kind of diurnal wave, and the superposition of cyclonic or anticyclonic heat-disturbance on its slopes, explains most satisfactorily what the author has so often observed in the United

States tri-daily weather-maps, viz. that the shape of the isotherms always appears to change more between the morning and afternoon than between the afternoon and night charts; and also that between the two latter, the shape often remained pretty constant, though the numbering had changed. For instance, in Figs. 42, 43, 44 we give

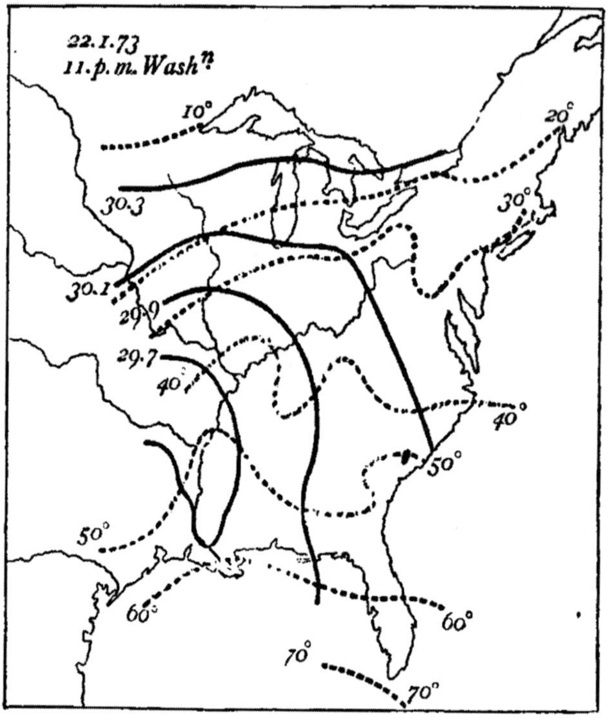

FIG. 42.—Diurnal and cyclone temperature (United States).

reductions of the United States charts at 11 p.m. on the 22nd of January, 1873, as well as those at 4.35 p.m. and 11 p.m. the following day. These are to serve a twofold purpose—first, to show why the distribution of temperature was so different on two consecutive days at the same hour, viz. 11 p.m.; and, secondly, to illustrate the diurnal variation in the shape of the isotherms between 4.35 p m. and 11 p.m. the second day.

We will consider the latter first. The isotherms which we see on the 4.35 p.m. chart (Fig. 43) represent the distribution of temperature due to the influence of a cyclone

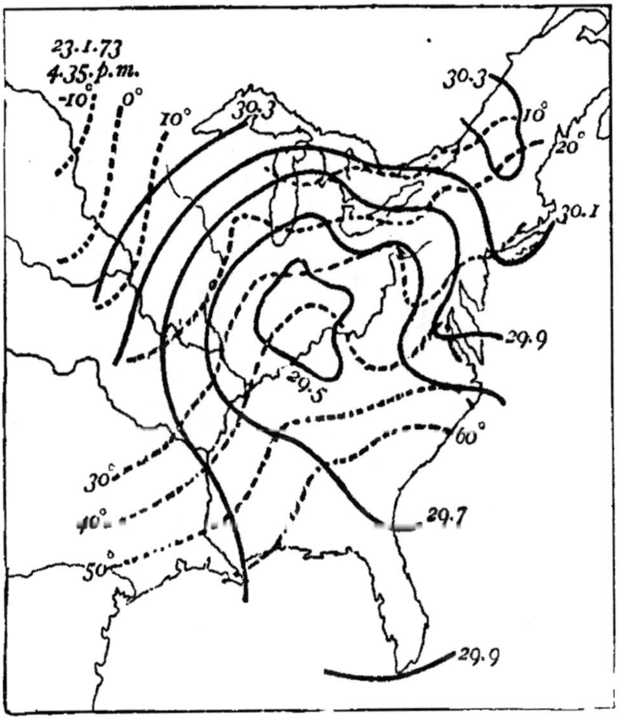

FIG. 43.—Diurnal and cyclone temperature (United States).

on a general irregular thermal slope from the equator to the pole, as modified by the diurnal range of the season. The aspect of the diurnal gradient is towards the north-east, because the temperature is falling.

By 11 p.m. the same day (Fig. 44) the centre of the

cyclone has scarcely moved, and the general shape of the isotherms is also nearly identical; but the position of the isotherms of 40°, 50°, 60° at 4.35 p.m. is taken broadly by those of 30°, 40°, 50° at 11 p.m., and the place of 10°, 20°, 30° at 4.35 is less nearly approached by those of

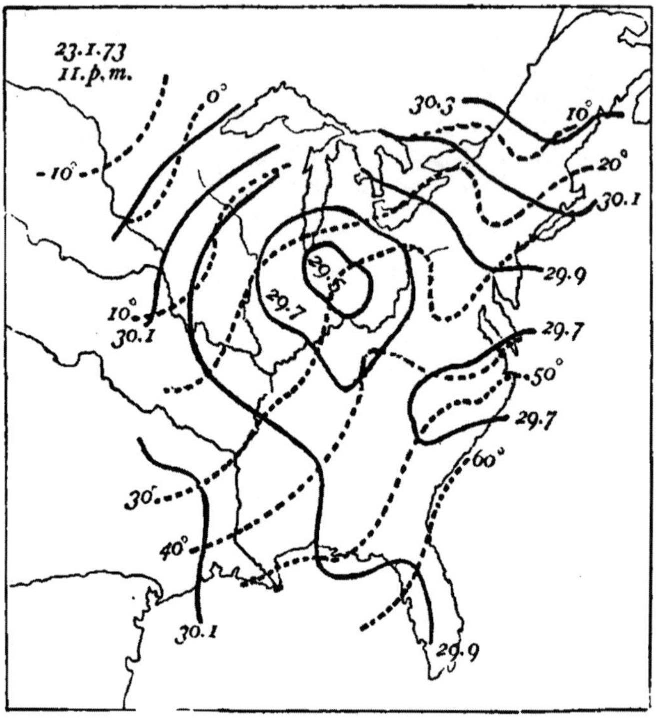

FIG. 44.—Showing diurnal and cyclonic temperature in the United States.

0°, 10°, 20° at 11 p.m. in the west. This means that the diurnal range was less in the north-west than in the south.

The interpretation of this is, that the aspect of the thermal gradients has not materially changed, though the temperature has fallen generally nearly 10°; so that the shape and position of the disturbance of temperature set up by the cyclone remains the same, but the numbering of the isotherms is changed nearly 10°.

Temperature-Disturbance of a Cyclone.

Now that we have eliminated the influence of diurnal range, we can better understand the nature of the heat developed by a cyclone. In the same three figures we have got rid of diurnal range by two methods. By taking the charts at the same hour—11 p.m.—in the first and third (Figs. 42 and 44), diurnal range is allowed for by being equalized, so that the whole of the difference between these two sets of isotherms is due to general changes, not to diurnal variations.

Then, by our second method of inferring the influence of diurnal slope on any shape of isotherms, we are enabled to use a chart at the intermediate hour of 4.45 p.m. (Fig. 43) for the same purpose of discovering the nature of cyclone-heat.

In all these charts we see that the general nature of the development of heat by a cyclone consists of a certain wedge-shaped projection of the isotherms northwards in front and on the southern side of the cyclone-centre, and that this heat moves on along with the cyclone. Observe that the local seasonal thermal gradient, from the cold interior of the continent to the warm sea, slopes to the north-west, while the aspect of the diurnal thermal slope is towards the north-east in all the charts. The quality of cyclone heat is very peculiar. It is not the pleasant warmth of a fine day, but has that characteristic close, muggy, disagreeable feeling which we have before described as coming before cyclones. This is the kind of heat which develops neuralgia and similar troubles in old wounds, and many of the prognostics which are associated with the front of a cyclone. We could not have a more striking instance of the necessity of adding a descriptive account to all instrumental records of weather. Neither a thermogram nor a synoptic chart can distinguish between one kind of heat and another.

The cause of this heat is obscure. The author has shown * that it is not altogether caused by that backing of the wind towards the south which precedes the rainy portion of a depression, and that the rise of temperature seems due to some peculiar property of cyclone-action.

In an ordinary whirl of dust or leaves we find the particles most compressed on the side where the directions of rotation and translation coincide; that is to say, if the whirl is against the watch-hands, and the motion in any direction, the compression is always on the right-hand edge of the eddy, looking towards the front.

If we reflect that a chart of cyclone-heat shows a wedge projection of the isotherms on a general thermal slope, we can readily understand how such a form may be analyzed into a detached patch of heat lying on a general thermal slope. We are thus led to the conception of a patch of heat developed by the cyclone, and moving about with it, like all the other characteristics of such a whirl.

For this reason, we often find exceptionally high

* Abercromby, "On the Heat and Damp which accompany Cyclones," *Quarterly Journal of the Meteorological Society, London*, vol. ii p. 274.

temperature on the north of a cyclone-path in the rare cases when the propagation of the depression is towards the west instead of towards the east, as is generally the rule. We shall recur to the importance of this fact, and give an illustration, in our chapter on Forecasting by Synoptic Charts.

But we may now describe in more detail the temperature-changes in the United States for the twenty-four hours to which the chart refers.

In the first map (Fig. 42) the centre of an irregular cyclone is near Memphis; the isotherm of 50° projects into this depression; the isotherm of 40° reaches nearly as far north as St. Louis, and all the Mississippi valley below that city is warm. By 4.35 p.m. the next day (Fig. 43) the cyclone has moved in a north-west direction to Indianopolis, and the isotherm which now projects most is that of 40°. Temperature has fallen all over the Mississippi valley, from the cold winds in rear of the cyclone. But what we have to notice most are the temperatures recorded at the Ohio stations, just in front of the upward projection of the isotherm of 40°, which were as follows:—Toledo, 19°; Cleveland, 27°; Pittsburgh, 31°. By 11 p.m. the same evening (Fig. 44) the centre of the cyclone had only moved a few miles, but that was sufficient to bring the stations just mentioned more within the range of higher isotherms than earlier in the afternoon. That is to say, the thermometer rose at all those stations between 4.35 p.m. and 11 p.m., although in an ordinary way we expect to see the mercury fall with the sun. The actual figures were—Toledo, 3°, Cleveland and Pittsburgh 2° each, higher than in the afternoon.

But while temperature has been rising in Ohio, many of the stations in the lower Mississippi valley have lost from 5° to 8° from diurnal causes. Other stations, such as Montgomery, Alabama, have lost no less than 13° from a combination of diurnal and cyclonic influences. A glance at the charts will enable us to see this at once, for, while the Mississippi stations are in the same portion of the cyclone at both hours, the latter station was in front of the cyclone at 4.35 and in rear at 11 p.m.

A few years ago, no explanation could have been given of this apparent anomaly of the air getting hotter as the sun went down; but now we see that it was due to the temperature-disturbance of a cyclone overriding the ordinary variation of diurnal influences. In our chapter on Meteograms we showed how similar changes would affect the trace of a thermograph; now, to complete a comprehensive view of the subject, we have illustrated the same phenomenon by the totally different method of synoptic charts.

Nothing could show better the extreme facility with which synoptic charts enable us to study temperature-changes in spite of diurnal variation. But just as it is not at all obvious at first sight how changes in the position of isobars are reflected in a barogram, so nothing but a good deal of experience will enable the meteorologist to see readily how changes in the position and shape of the isotherms would affect the indications of a single thermometer, or to handle with any ease the idea of the propagation of diurnal isotherms over a complicated system of temperature-distribution. Our example is one of the simplest which the author could find. In the first and third charts we equalize diurnal influences by constructing the maps at the same hour each day. By this means we can explain why the Atlantic states were colder on the first day than on the second, and why the Mississippi valley was colder on the second than on the first day.

By our second and third charts we illustrate the manner in which general changes override diurnal variations when the latter are not very strong, as during the winter months, as well as the characteristic nature of a diurnal fall of temperature on an existing system of isotherms.

There are nearly eighty stations in the United States and Canada. During the six and a half hours in question, changes in every direction of varying magnitude took place at each; but there is not one, however apparently anomalous, which cannot be explained by means of the principles which we have here laid down.

Sources of Heat.

We were obliged to introduce the question of diurnal variation of temperature in the first place, so as to get rid of any ideas of difficulty from that source of complication; but now, before we describe further the changes in the isotherms from day to day, we must consider the various sources of heat and cold with which we have to deal. In all this we must never forget that the natural distribution of temperature is an irregular thermal slope from the equator to the pole, and that what we have to explain are the divergences from that ideal distribution which we find in practice. We shall find that places far north are sometimes much warmer than others nearer the equator, and that some parts of Europe are often colder in March than in January. All these apparent anomalies we can explain easily, but we must begin with sources of heat.

The primary source of heat is, of course, the sun, so that, other things being equal, we should get the greatest heats where there is the least cloud; that is to say, generally in anticyclones. This, however, cannot be laid down as a general rule without some modifications. In the belt of anticyclones which surround the world about the line of the tropics, some of the greatest known heats are recorded, notably in the Sahara and in Australia. But in higher latitudes the sun has a powerful enemy in the cold space which surrounds the earth. In summer the sun is the more powerful, and we get hot days with cold nights. In winter-time, when the sun is low, radiation into space overpowers the radiation from a low sun, and clear weather is cold. When, then, we come to discuss in general terms the influence of cloud on isotherms, we must always take into consideration the time of year and latitude.

Another very powerful source of irregular isotherms is found in wind. Of course, speaking broadly, southerly winds will deflect the isotherms northwards, and northerly or easterly winds will bend them towards the south. This too is, however, subject to many irregularities. The great difference in the radiating power of land and water at different seasons, makes a continental area colder in winter and hotter in summer than a sea in the same latitude. For this reason an easterly wind from a land-area would blow warm into a neighbouring sea in summer, and cold in winter.

We have already alluded to the idea of a specific quality of heat which is developed by cyclones, and minor local sources of variation, such as the descending winds which probably constitute the "fohn," need only be mentioned here.

When all these sources of heat are combined, a vertical sun, a cloudless sky, a southerly wind, and an arid soil; when light, hot puffs fill the air with scorched particles of sand, till the dulled sun appears to glow in a sea of molten brass, and the poisoned breath of the *simoon* sweeps fitfully across the desert, then the traveller may well beware, and hasten for his life to the nearest shelter.

An example of great heat will be found in the charts, Figs. 82 and 83, which illustrate the first burst of the south-west monsoon in Hindostan. The dates, June 17 and 18, 1881, would coincide with the end of the hot season in Northern India. Our diagrams show on both maps a patch of heat over 100° Fahr. (38° C.) over the Desert of Scinde; and, as this would be at about half-past five in the afternoon locally, it is certain that much higher temperatures must have been recorded nearer midday. The isobars show that what wind there was would be southerly or south-westerly, and of course light from the absence of gradient near the centre of the depression. The soil there is saltish sand, and similar material has been known to get heated up to nearly 200° in Australia. In Scinde there is a dangerous wind at this season exactly analogous to the simoon of Arabia and the Sahara. Both are certainly allied to whirlwinds and tornadoes; but, unfortunately, no scientific observations have been made on the reputed poisonous or fatal character of these blasts, or of the dangerous quality of heat which they develop.

There is one type of warm weather in Europe for which no explanation can be given at present. We have seen that an anticyclone usually develops cold-radiation weather, but sometimes we find an anticyclone with warm air and a peculiar soft cloudy sky. This anticyclone covers Continental Europe, and is always associated with the eastward passage of distant cyclones on the northern side. No reason can be assigned for this heat; all we can do is to note the fact for future research.

Sources of Cold.

The principal source of all cold is radiation into space. The space which surrounds the earth has a theoretical temperature of at least 226° below 0° Fahr., and it is the influence of this chilly envelope which we feel.

The greater part of the influence is, however, indirect. We do not feel the cold of space as if we were standing near an iceberg, for all our greatest colds are produced by radiation. Bodies on the earth's surface radiate into this cold space till they lose a large amount of their original temperature; and air, which is a bad radiator itself, gets cold by contact with the chilly soil.

For instance, on a calm winter night different bodies—say, a sheet of iron lying on the ground and a patch of grass—begin to radiate into space at different rates, according to their own intrinsic properties. Iron radiates very quickly, but is also such a good conductor that it brings up an abundant supply of heat from the ground to replace the loss by radiation, so that the plate does not become very cold. The grass is a less good radiator, but at the same time a very bad conductor; so, though it parts with heat slower than the iron, it cannot replace what it has lost by conduction, and therefore, on the balance, becomes much colder. This is the cold which we really feel, and which sends down the thermometer.

A very striking result of all this is, that under these circumstances, the air gets warmer as we ascend up to a certain height, and this proves conclusively that we do not feel directly the chill of space. Of course, the greatest cold will be produced when the greatest number of causes are combined which favour radiation. These are a still air, a clear sky, and an absence of water-vapour in any stratum of the atmosphere. This last condition is very interesting. Professor Tyndall's researches seem to show that water vapour is a great absorbent of the quality of heat which is radiated from the ground, so that when much vapour is present the ground cannot lose its heat so rapidly as when the air is dry.

All these conditions of great cold are fulfilled in the most perfect manner in Siberia. There we have the centre of a large dry continental area, which in winter-time is persistently covered by an anticyclone; while the latitude is so high that the sun has little power. Here, then, we find calm, dryness, and a feeble sun; and here the greatest known colds are reported, if we except some in the north of Smith's Sound, many degrees further north. A good illustration of this will be found in the two charts which we give in our chapter on types of the north-east monsoon (Figs. 80 and 81). In them we see that the south of Siberia, which is covered by an anticyclone, has stations in which the mercury marks more than 30° below zero, Fahr.

If we take the less extreme cases which occur in Great Britain, we find that all frosts in that country are "home-brewed;" that is to say, that cold winds never bring extremely low temperatures from the plains of Europe or the mountains of Norway. But when shallow gradients for east and north-east winds cover Great Britain, and a dry chilly air favours nocturnal radiation, then all the hardest frosts are developed. Then we often find the temperatures 10° or 20° lower in the most inland stations of England and Ireland, and the isotherms gradually increase round these cold centres. When we look at a synoptic chart of Europe for 8 a.m., we find, on these occasions, that England and Ireland are separate islands of cold on the general thermal slope from a cold continent to the warm North Atlantic. From the fact that frost depends on radiation, we can readily explain why cold is so local. Radiation is very sensitive; the least breath of wind or any local shelter may interfere with the free play of radiation, and so we find two places only a few miles apart, one of which records 10° or 15° lower than the other.

The next source of cold is found in wind. When this blows from a frozen continent, then, of course, very low temperatures may be recorded; but this is not the same kind of cold as radiation-frost. Here we have another of the innumerable instances of the necessity of distinguishing between different kinds of the same nominal phenomenon. The 1st of January may be cold in one year from wind; in another from radiation. These are the products of totally different kinds of weather, and must not be mixed up in scientific meteorology.

THE "BLIZZARD" AND THE "BARBER."

A very striking example of wind is found in the "blizzards" of the United States. These are cold snaps which come with a high wind, as opposed to the calm frost of anticyclones. They are the result of the passage of the rear of cyclones or of V-depressions in the winter months, such as we see in Figs. 42 and 43. Then we get high, strong, north-westerly winds, blowing off a frozen continent with a temperature many degrees below zero, and with surroundings which are very destructive to life. The wind drives the cold into the bones even through fur clothing, and raises a blinding dust of powdery snow. Under these circumstances only are the western voyagers ever lost. If wood cannot be found, nature can only resist the cold for a certain number of hours, and the men are frozen to death if no shelter can be reached. A very curious circumstance attends these deaths. In almost every case the victims are found to have begun to strip themselves. When the body is nearly reduced to an icicle, only a very little blood continues to circulate languidly through the brain. Then delirium sets in, with a delusive sensation of heat, under the influence of which the traveller begins to divest himself of his clothes.

Another disagreeable form of cold is found in the

St. Lawrence Gulf. Sometimes with a high wind the air becomes much colder than the open water. The latter, being relatively hot, begins to smoke, and the vapour freezes into peculiarly sharp spicules. The *poudré* snow-crystals of the north-west are usually small, dry, six-sided petals, and, though penetrating as sand, they are soft. The latter kind of snow is so damp and sharp that, when driven by a gale, it nearly cuts the skin off the face. Hence the popular name of the "barber," which is applied to this phenomenon. The same name of "barber" is applied to another phase of cold along the coasts of Nova Scotia and New England. When a vessel is caught by a gale of wind in a cold arctic current, the spray freezes the moment it touches the deck or rigging. Every block is turned into a lump of ice; men get coated with ice like an icicle; and sometimes such a weight of ice forms on the bow that the stern is lifted out of the water, and the ship becomes unmanageable for want of steering power.

The last source of cold which we need mention is rain. All rain, of course, is not cold. In front of a cyclone rain is warm, and a shower does not send down the thermometer. In the rear, on the contrary, and in thunderstorms and secondaries, precipitation is more or less cold, and turns the mercury downwards. The influence of this varies very much in different countries and at different seasons of the year. In England, during the summer, rainy weather is cold, because it cuts off the sun, independent of any chill of its own. In winter, on the contrary, rainy weather is warm, because an overcast sky prevents loss of heat by radiation. In the tropics cloudy weather is colder, as far as the thermometer is concerned, than a bright day, because the rays of the sun are obstructed; but if there is little wind, a cloudy day is more oppressive to men than one with sunshine. Near the equator there is very little diurnal radiation of any kind, owing to the excessive amount of vapour in the air.

We may sum up all the effects of heat and cold briefly thus: In winter wind, cloud, and rain in temperate regions tend to raise the temperature, as they check cold radiation; calm, on the contrary, induces hard frost. In summer wind, cloud, and rain are cooling influences, as they check hot radiation; calm, on the contrary, is then hot, because it allows full play for the sun's rays.

We may, in fact, look at the opposing forces of hot and cold radiation as in a state of constant conflict. The rotundity of the earth always weakens the power of the sun in the north. Water-vapour in some shape forms, as it were, a blanket for the earth, and saves her from being burnt up and frozen alternately. The incessant circulation of the atmosphere sometimes eddies in a cyclonic form, and develops dense cloud, which shields the earth from the radiation of the season and latitude; at other times the circulation of the air eddies downwards in an anticyclone, and the clear, dry, calm atmosphere gives full play to radiation, and some extreme of heat or cold is then developed.

The task of the meteorologist is to trace how the varying forms of atmospheric circulation modify the distribution of heat and cold over the world from day to day, by the application of the general laws we have just laid down.

Examples of Daily Temperature-Changes over Europe.

For instance, let us consider the changes of temperature which occurred in Europe on the three days, February 26–28, 1865; that is, during three of the days for which we shall give synoptic charts in Figs. 68–70, when discussing the westerly type of weather. We have to explain now why European temperature varied as it did on those days.

The isotherms for the period in question are given in Figs. 45-47. In all of these there is a thermal gradient

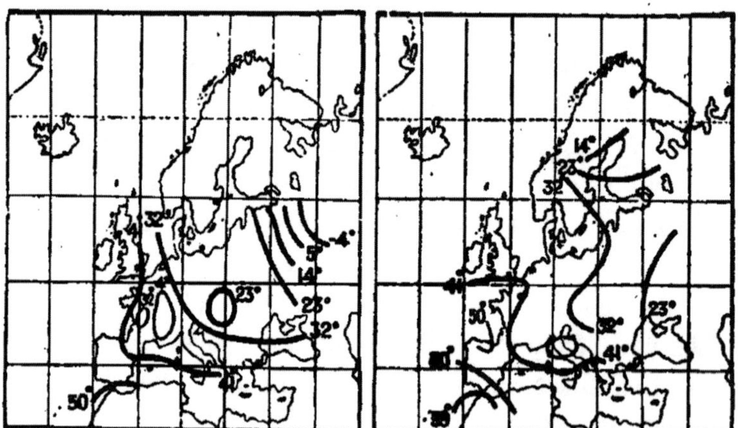

Feb.26.1865, 8 am.Greenwich. *Feb.27.1865, 8 am.Greenwich.*

Figs. 45 and 46.—Isotherms in Europe for three consecutive days.

from south-west to north-east instead of a slope towards the north-west, as undisturbed natural isotherms should have at eight o'clock in the morning—the hour for which the charts are constructed. The reason for this broad feature is that in winter a continental area is always colder than a sea-surface, and therefore, whatever smaller variations may occur from day to day, the general slope of temperature will always be from frozen Russia towards moisture-bathed Portugal. This feature belongs to the season, and is found in every chart; what we have now to explain is the fluctuation in the position of the isotherms caused by the varying development of heat and cold locally.

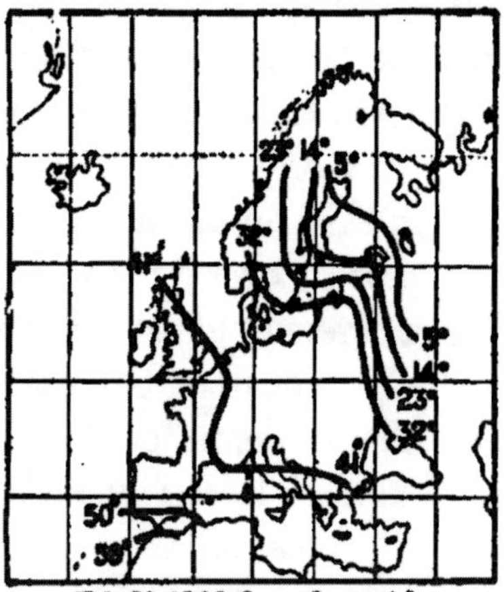

Feb. 28. 1865, 8 am. Greenwich.

Fig. 47.—Isotherms in Europe for three consecutive days.

Glancing at both the synoptic charts of pressure and temperature, we see that on the morning of February 26 a V-depression covered Great Britain, with warm south-west winds in front. Straight isobars lay over Scandinavia, an anti-cyclone stretched over Western Europe from the Atlantic, and a calm col lay over Russia. From all this England was warm, as shown by the projection northwards of the isotherm of 41° Fahr. (5° C.); Continental Europe and Russia were very cold. In the latter country, −4° Fahr. (−20° C.) is reported, and local patches of cold as low as 4° Fahr. (−15° C.) are reported in different parts of France and Germany. These should be noticed, for they are most characteristic of the abrupt local variations of temperature which we often find are caused by local differences in radiation. They are identical with all the frosts which occur in Great Britain, to which we have before alluded. Observe also that they have no influence whatever as a cause of weather; they are the product of the general circulation of the atmosphere, allowing free play for radiation, not a cause of that circulation themselves, though the influence of the general thermal slope from Russia to Portugal is an important factor in determining the path of the cyclones. By next morning the British V and Scandinavian straight isobars have formed a well-defined cyclone, some secondaries appear over various parts of Europe, while a calm wedge covers England. England is, therefore, colder than on the previous day, because of the radiation of the wedge, and the isotherm of 41° Fahr. (5° C.) has retreated southwards. Russia and Continental Europe are much warmer, because the cyclones and secondaries have destroyed radiation.

By the morning of the third day the Scandinavian cyclone has died out, but a new one lies over the north of Scotland. Secondaries still cover the greater portion of Europe, but in Russia the weather would be calmer. From this it results that England is warmer, so that the isotherm of 41° Fahr. (5° C.) projects northwards again; Continental Europe is a little colder, without many local frosts; Russia is a great deal colder, but not so cold as the first day, for the conditions are less favourable to radiation. None of these cyclonic changes reach so far south as Spain, and therefore we see the isotherm of 50° Fahr. (10° C.) scarcely alters its position during the three days.

We may also put the changes of temperature over Europe in a very striking light by looking at the isotherm of 32° Fahr. (0° C.). On the first day it stretches from Belgium to the Black Sea; the second day it has been driven back almost to the Gulf of Bothnia and to Poland; the third day it has advanced again, but not so far as on the first day. So on the conflict would go between the frost and sun till the sun at last drove that isotherm out of Europe. In the autumn the battle would be renewed; but then the sun would be beaten, and frost remain supreme for several months in the more northern portions of that continent.

Had our limits permitted, we would have given examples of the reversal of radiation effect which occurs in summer, when an anticyclone means heat instead of cold. Then we may often find England hotter than France, for if the calm centre of an anticyclone lies over the former country, the sun's rays have more power there than in the more windy southern edge, which would cover France under these circumstances.

We may, however, refer again to Figs. 21 and 22, which relate to the same day of May—the 17th—in two different years, and in which diurnal variation is allowed for by constructing the charts at the same hour. On the first day (Fig. 21) a cyclone covers Great Britain, and the isotherm of 50° Fahr. (10° C.) reaches to the north of Scotland and Denmark, under the influence of southerly winds and a cloudy sky.

On the second day (Fig. 22) the isotherm of 50° Fahr. (10° C.) runs north and south down England, and a corner of the line of 40° Fahr. (5° C.) appears over Northern Germany. This shows that to the west of the isotherm of 50° the temperature rises towards 60° Fahr. (15° C.), and therefore that part of England and Ireland is warmer than on the same day of another year, when no place recorded anything as high as 50°. This is the product of the calm blue sky of an anticyclone; while the diminished temperature over Germany is due to the general thermal slope of the season, for Continental Europe does not get warm till the month of June.

If we combine all these with the other examples we have already given of temperature ranging from 100° Fahr. (40° C.) to –30° Fahr. (–35° C.) in Europe, Asia, and America, the reader will have a very fair idea of the nature of temperature-changes.

Forecasting Temperature.

From all this it will be very evident that, though we can lay down some general laws of temperature-changes, still the modifications which occur in practice are endless. The forecaster in every country has to learn by experience the qualities of the different winds, and the power of the different radiations at each season of the year.

For instance, in Great Britain he soon learns to distinguish between the uniformly warm, close heat of winter cyclones; the oppressive, sultry heat of a summer thunderstorm; and the clear, cold air, with a hot sun, of a spring anticyclone. Any doubt which can arise as to the future course of temperature-changes depends on the same points which always make any forecasting uncertain —viz. the difficulty of knowing what the future path of the cyclones will be, or whether any new distribution of pressure is likely to set in suddenly. If the forecaster judges rightly as to the future movements of pressure-distribution, he rarely makes a mistake as to the nature of temperature-changes which accompany them.

Primary and Secondary Effects of Heat.

We will conclude with one important reflection. We know that heat is the prime mover of all atmospheric circulation; why, then, do the great local differences of temperature have so little influence on the sequence of weather? The greatest diurnal ranges are found in anticyclones, which are also associated with the steadiest weather; and in wedges, where we find strong contrasts of heat and cold, these local differences of temperature are certainly not the cause of the cyclone and rain which follow soon. At the same time, it is certain that the persistent anticyclone over Siberia during the winter months is caused by the radiation cold of that country. That is to say, we may conceive that in the general circulation of the hot air of the equator towards the pole, the direction of the currents will be profoundly modified by the surface-temperature of the earth, and that it is perhaps easier to flow over a cold surface at one season and a warm one at another.

However that may be, we are met by the apparent contradiction that, though the daily variations of temperature are undoubtedly the product of the motion of cyclones, etc., the broad situations of the areas of cyclone activity are themselves due to radiation.

The truth probably is that both inferences are correct in a modified degree, and that in this, as in every other meteorological problem, we have to deal with a balance of influences which act and react on one another in a very complicated manner.

We have already explained the stability of a circulatory system such as a cyclone or anticyclone, and the idea that diurnal variations may merely affect the rapidity, but not the form, of the vortex system; but one observation may perhaps be noted here which probably has some bearing on the question. Our synoptic charts give surface-temperature only, but we have taken no notice of the heat of upper currents. Now, it has been discovered that over cyclones temperature diminishes from the surface upwards at the rate of about 3° Fahr. (1·5° C.) in one thousand feet; in anticyclones, on the contrary, when radiation produces frost, the air gets warmer as we ascend to a short distance, after which the temperature begins to fall as we go higher up.

What the precise significance of this may be we cannot tell, but it is interesting in connection with the manner in which pressure decreases at a slower rate over cyclones as compared with anticyclones. Further research can alone solve these problems, to which we have merely alluded to carry out our purpose of giving a picture of the state of meteorology at the present day.

CHAPTER VIII.

SQUALLS, THUNDERSTORMS, AND NON-ISOBARIC RAINS.

IN this chapter we propose to introduce the reader to details of weather totally different from any that we have hitherto described. So far we have dealt with the phenomena of wind and rain which are associated with cyclones and rapid changes of barometric pressure; now we intend to discuss changes of weather which are connected but indirectly with the distribution of surrounding pressure, and in which, if the mercury moves at all, the direction is upwards. Isobars which have been our unerring guide through the most complicated cyclonic weather will now totally fail us; and, under the heading of non-isobaric rains, we shall discuss certain rainfalls, to the origin of which we have at present but little clue. In addition to the interest which attaches to such striking manifestations of nature as squalls, thunderstorms, tornadoes, and whirlwinds, a great deal of research has been bestowed of recent years on these subjects which has not yet found its way into popular literature, and which at present is scarcely known beyond the limited circle of professional meteorologists. We now propose to explain some of the most remarkable results which have thus been obtained.

SIMPLE SQUALLS.

If we watch the stages of gradually increasing wind, we find that as the strength rises the tendency is more and more to blow in gusts. Gradually these gusts get still more violent, and in their highest development come with a boom like the discharge of a piece of heavy ordnance. This is what sailors call "blowing in great guns," and these are the gusts which blow sails into ribbons, and dismast ships more than any amount of steady wind. These gusts only last a few minutes, but they seem to be very closely allied to the simplest form of squalls. In a true, simple squall the wind generally need not be of the exceptional violence which causes "guns;" but after it has rather fallen a little, the blast comes on suddenly with a burst, and rain or hail, according to intensity, or other circumstances, while the whole rarely lasts more than five or ten minutes. At sea one often sees two or three squalls flying about at a time. Then we readily observe that over the squall there is firm, hard, cumulus cloud; that the disturbance only reaches a short distance above the earth's surface; that the squall moves nearly in the same direction as the wind; and that there is little or no shift of the wind before or during the squall. We also see that the shape of the squall is merely that of an irregular patch, with a tendency rather to be longer in the direction of the wind than in any other quarter; and that the motion of the squall as a whole is much slower than that of the wind which accompanies the first blasts. If, at the same time, we watch our barometer closely, we find that if the squall is sufficiently strong, the mercury invariably rises—sometimes as much as one-tenth of an inch—and returns to its former level after the squall is over. No difference is observed in this sudden rise, whether the squall is accompanied with rain, hail, or thunder and lightning; and though we are unable exactly to explain why the wind sometimes takes this irregular method of blowing, we have still to do with a comparatively simple phenomenon.

THUNDER-SQUALLS.

The simplest kind of thunderstorm may more properly be described as a squall accompanied by thunder and lightning, instead of only with wind and rain. In Great Britain these thunder-squalls are very common on our extreme west and north-west coasts during the winter months, while they are very rare in Central or Eastern England at any season of the year. On a wild, stormy day, with common squalls, one or two of these, which are exceptionally violent, will be accompanied by one or two claps of thunder with lightning. The principal interest which attaches to this type of thunderstorm consists in the proof which is afforded that there is no essential difference between a common squall and another which may be associated with electrical discharge, except intensity. The look and motion of the clouds, and the sudden rise of the barometer, are identical in both cases.

We can readily conceive, since the formation of cumulus above the squall points unmistakably to the presence of an ascensional current, that when the uptake is only moderate, the condensation of vapour may take place so gradually that none of the electricity—which there is reason to believe is given off under these circumstances—is discharged disruptively; but when the uprush is so violent as to inject the moist air into strata which are so cold and dry that the electricity cannot pass off silently, then a disruptive discharge with thunder and lightning will be produced. In Western Europe this class of thunderstorm is much more common in winter than in summer, which is the reverse of what takes place with all other kinds of thunderstorm. So much is this the case that in Iceland there are no summer thunderstorms, but only winter ones, of this simple squall type. In Norway both types occur; and the winter ones are there found to be the most destructive, because they are lower down, and therefore the lightning is the more likely to strike buildings. In that country, however, the summer thunderstorms are not nearly so violent as in more southern latitudes.

Barometer in Squalls and Thunderstorms.

We have just mentioned that the barometer usually rises just as the rain of a squall or thunderstorm strikes a place, and this is as true on the Equator as on the Arctic Circle. Since this fact is of great importance in the discussion of the more complicated phenomena that are called line-squalls, we will devote a few paragraphs to the elucidation of the details of these barometric fluctuations. We can do this best by an actual example.

In Fig. 48 we give a photographic engraving of the barometer-trace given by the author's barograph in London, on May 18, 1878. The original was recorded on smoked paper, and is here reproduced by photography, absolutely untouched by the engraver. By this means the most delicate fluctuations are faithfully rendered, and those who are familiar with sensitive self-recording instruments will readily recognize that characteristic uneasiness of the whole trace, which can never be copied by hand. In the figure the vertical lines represent intervals of six hours, while the horizontal lines indicate a difference of 0·5 inch of mercury. Confining our attention to the right-hand portion only of the diagram, we have to note, soon after midnight of May 17, a small curious dip of the barometer, followed immediately by a sudden rise.

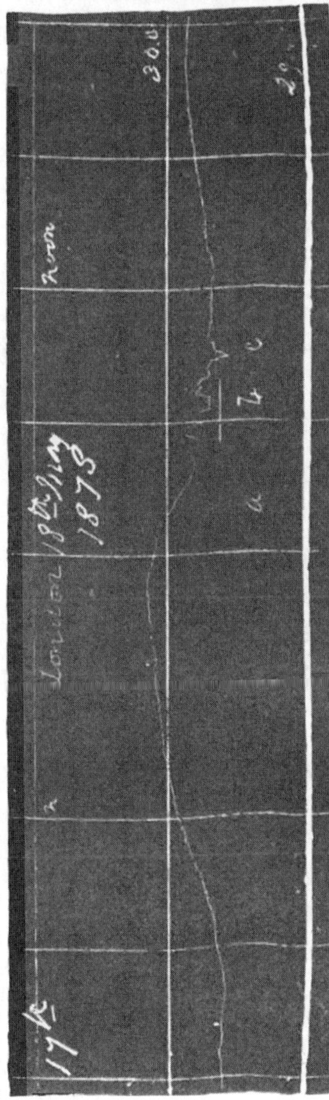

FIG. 48.—Barometer in thunderstorms.

This is marked *a*, and it occurred during a thunderstorm. Just before 6 a.m., and for some time after, we find the still more remarkable fluctuations marked *b*. These were also associated with a series of thunderstorms, none of which were particularly violent. Still later, about 8 a.m., we see the singular dip marked *c*. This occurred with gloomy, threatening weather; but neither with wind, rain, nor thunder, at the place of observation in London.

The chart for the day at 8 a.m. showed that a series of small secondaries lay over Great Britain, but there were no bends in the isobars that would explain such curious barometric oscillations.

The origin of this characteristic rise of the barometer in squalls and thunderstorms is at present unknown. It has been suggested that it is due to a rush of air, carried down by the rain. That such is partially the cause is extremely probable, for we sometimes see a small rise under a heavy splash of rain without either thunder or wind. But it is equally certain that this downrush does not entirely explain the phenomenon, for sometimes a rise occurs without any rain at all, or of an amount which bears no relation to the heaviness of the fall. Still more puzzling are the small dips of the mercury which we occasionally find with thunderstorms, and of which some examples are given in our last figure. These dips are more rare than the rises, and though in most cases they are, as in this example, more or less associated with the rises, still they occasionally occur alone. In the first dip shown in our last figure, about 1 a.m., the depression was associated with a storm; while in the second case, about 8 a.m., no storm or rain occurred locally, though undoubtedly storms were in existence not far off. We are, therefore, almost driven to the conclusion that some of these curious fluctuations of the barometer must be due to a sort of true wave-action, through which the disturbance, caused perhaps by falling rain, may be propagated by the elasticity of the air to some distance from the place of original disturbance. In connection with this idea of air being brought down by falling rain, we may notice that very striking effects are sometimes observed in avalanches of snow, which always bring down an immense amount of imprisoned air with them. It is usually found that persons caught in the blast of the avalanche have their clothes torn into ribbons. The suggestion has also been made that if rain is the product of the condensation of an ascensional current of air, then the more violent the uptake, the greater must be the reaction downwards; but, unfortunately, our knowledge of the dynamics of air in motion is not sufficiently advanced to enable us to say exactly what the nature of pressure would be under these circumstances.

But though we cannot altogether explain the origin of these barometric fluctuations, we know enough to say that they are of a totally different nature from any motion of the mercurial column due to the action of cyclones or the propagation of isobars over any station. When, then, we see on a barogram these peculiar irregularities, we can at once infer that they are the product of squalls or thunderstorms, and not of cyclones, and so far we are enabled to increase our knowledge of the method of reading barograms, to which we have already given so much attention. These dips and rises may, in fact, be taken as another letter in that barographic alphabet by which a skilled meteorologist can read the history of the weather from his barometric trace. Another very important inference which this knowledge gives us is that, as these rises are entirely different from those due to cyclonic motions, we cannot make the same deductions from the one that we would from the other. For instance, if in rear of a cyclone the mercury rose at the rate of two-tenths of an inch in an hour, that would be an exceptional rate anywhere in Europe, and we should expect that it would be associated with a violent gale. But the mercury might rise at two or three times that rate in a thunderstorm of no exceptional intensity, with which there would be no more than a few irregular gusts. For the same reason, if we have to include any barometric readings of these peculiar rises in the construction of our synoptic charts, we must not draw the same conclusions from the lie of the isobars as we should in ordinary weather, because the origin of the isobars is not the same in both cases. The error which we have to avoid is not to take as the same two phenomena that are really totally distinct, but which have one common property—namely, a rise of the barometer.

Line-Squalls.

We have already explained that the line of the trough of a cyclone or V-depression is associated with a line of squalls, and that we must picture to ourselves a long, narrow, thin band of rain and wind sweeping across the country, broadside on, like a wall or curtain, at the same speed as the depression itself. This rate bears no relation to the velocity of the wind in the squall. In practice the rate of the depression will be much slower than the wind in the first gust. The former will probably not exceed forty miles an hour; the latter may mount up to seventy or eighty miles an hour. We will now go into some very interesting details of this class of atmospheric disturbance, which, for the sake of classification, we will call "Line-squalls."

The nature of this class of squall will be best explained by an actual example of the squall which capsized an English man-of-war—the *Eurydice*—and caused one of the greatest disasters which has befallen the British navy for many years. In Fig. 74, under weather-types, we give a chart of a large portion of the northern hemisphere for March 24, 1878, at 0.43 p.m., Greenwich, and we may just glance at it now to see the general distribution of pressure over Europe on that day. The squall which we have now to consider belonged to one of the numerous secondaries, which hardly show on the large chart; but in Fig. 49 we give the details of pressure, wind, and weather over Great Britain and France at the same hour on a larger scale. In this diagram we see an extremely complex distribution of pressure. What concerns us most is the bend in the isobars, along which we have run a dotted line that is marked "trough" at one end. This bend is a small V-depression, in some way secondary to the ill-defined fragment of a cyclone that covers the southern portion of the Scandinavian peninsula. During the course of the day this cyclone appeared to circle round another which lay in the morning over the Carpathian

Mountains; and, in connection with these greater changes, the trough of the V wheeled round a point near the Scaw, in Denmark, like the spoke of a wheel. Fig. 49 shows the position of the trough at 0.43 p.m.; the front line of the crescent-shaped shaded area in Fig. 50 shows approximately the position of the trough at 3 p.m.; and by 6 p.m.

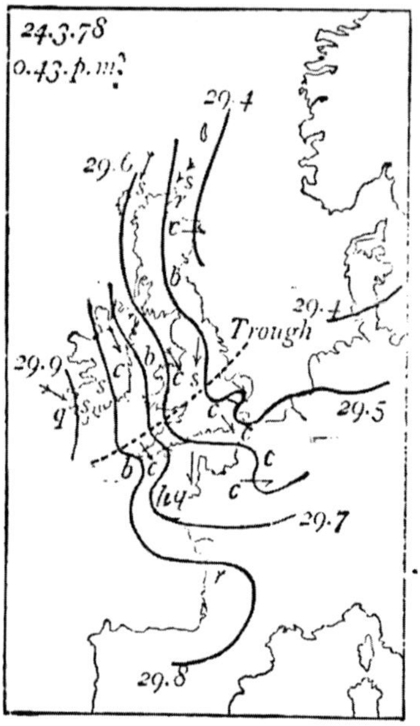

FIG. 49.—The *Eurydice* squall. Isobars and wind at 0.43 p.m.

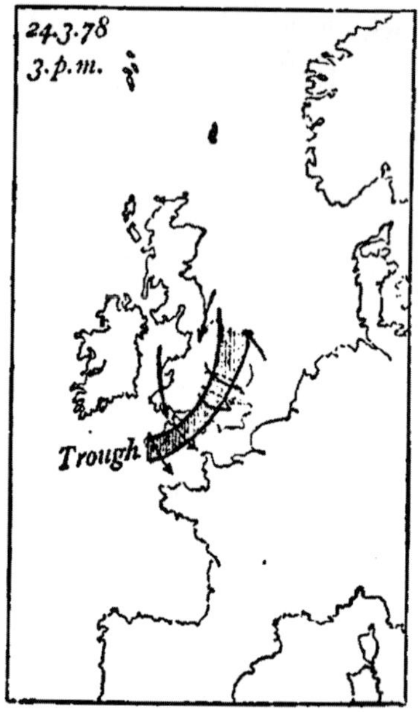

FIG. 50.—The *Eurydice* squall. Area covered by squall at 3 p.m.

the trough passed in a curved line from Yarmouth, through the Straits of Dover, into Normandy. By reason of this wheeling motion, different portions of the trough moved with very different velocities. Between the hours just named, the northern portion of the trough moved across England at the rate of only thirteen miles an hour, while the extreme south-westerly edge traversed the country at the rate of no less than forty-eight miles an hour. The portion which struck the *Eurydice* was going at the rate of thirty-eight miles an hour.

So far for the motion of the V as a whole. In Fig. 49 the wind was from about west in front, and from north-west in rear of the V, but no well-defined area of rain was then developed. By 3 p.m., however, the depression had so much increased its intensity that Mr. Ley was able to construct the diagram given in Fig. 50. In that figure the shaded portion shows the area over which rain or snow was falling at the moment; the solid arrows give the general sweep of the surface-winds, the dotted ones those of the upper currents. The author has further shown that the front of the rain-area was essentially coincident with the trough of the V, which we see about three hours earlier in Fig. 49, so that we evidently have to deal with a V of that class in which the rain is in rear of the trough. At every station, after the wind had been from the west, with a cloudy sky in the morning, the clouds gradually banked up ominously to the north-west; then rain or snow came on with a tremendous squall, while the wind jumped round to north-west. After the first burst had moderated, rain or snow continued for a longer or shorter interval till the sky cleared again.

H.M.S. *Eurydice* was a full-rigged corvette, homeward bound from the West Indies. At 3.45 p.m.—three-quarters of an hour later than that for which the chart given in Fig. 50 was constructed—she was off Ventnor, in the Isle of Wight, running free before a nearly westerly wind, with all sail set. At that moment she was struck by a squall from the north-west; before sail could be shortened, she went on to her beam-ends, and, as the lee ports were open, she filled and foundered.

On the whole we have, therefore, to idealize a band-shaped area of rain, bounded in front by a line of squalls —in this case more than four hundred miles long—sweeping broadside on across Great Britain at a rate varying from thirteen to nearly fifty miles an hour. From this we can readily see how places many miles apart can be struck simultaneously at the same hour, and how applicable is the name of line-squalls, which we have applied to this class of disturbance.

Though this class of line-squall is uncommon in Great Britain, it appears to be very frequent in other parts of the world. For instance, in Iowa, a similar kind of squall is peculiarly characteristic of summer weather. There it generally occurs after a spell of continued hot, rather sultry weather, the wind having blown steadily but moderately from the south or south-west, the barometer not changing much. In the north-west the storm-front will make its appearance; threatening, dark, towering clouds, or at times an immense roll-like cloud, will approach; the air cools rapidly as the storm-front comes nearer; and, with a high, straight blow, bending young trees to the ground and driving the rain nearly level, the fierce storm passes over, while the barometer rises rapidly. Such a blow does not last long, but may be repeated with gradually weakened force at intervals. A steady pouring rain generally follows, after which the sky clears, and the storm-wind wheels back to the south-east, the weather being as hot as before the storm. This description, which we have taken from Dr. Hinrichs, of

Iowa City, together with the maps which he gives to illustrate them, point very clearly to line-squalls, associated with that class of V-depression in which the rain follows the trough. His maps do not exhibit the shape of the rain-area like the one we have just given, but they show the squalls sweeping across the State with a crescent-shaped front exactly like the *Eurydice* squall.

When we compare this class of squall with the pure and simple squall which we first described, it will be obvious that the two kinds have little in common except the name. The former kind seems to be simply a local intensification of a general sweep of wind. The latter, on the contrary, is associated with a very definite but complex phase of aerial circulation, which we shall understand better when we have described a precisely analogous class of thunderstorms.

Thunderstorms associated with Line-Squalls.

Squalls are rarely of sufficient importance to attract the notice of enough observers to enable the details of their shape and progress to be properly determined; but thunderstorms are such a striking manifestation of weather that they are much more easily traced, and an enormous amount of work has been done in late years in marking out the hourly advances and development of such disturbances.

Many, but not all, European thunderstorms have been found to be precisely similar to the line-squall which we have just described. Some of Bezold's diagrams of

Bavarian thunderstorms, which give the shape of the area covered by the storm at successive hours, show long narrow bands sweeping broadside on across the country exactly analogous to the squall-area which we drew in Fig. 50.

But the remarkable point is, that though some of these storm-bands are associated with the troughs of cyclones and V's, precisely similar bands are more often found either in front or in rear of the cyclone, where we can connect them (the bands) with no particular part of the cyclone, except that the front of the band is usually perpendicular to the line of progress of the depression in which it is formed.

We will first give an example of the storm and thunderstorm associated with the trough of a V-depression. On July 16, 1884, at about 6.15 p.m., the trough of a V-depression swept over Hamburg, and brought, as usual, a violent squall and heavy rain, with much thunder and lightning. This was only a section, as it were, of a line-thunderstorm. Dr. Sprung, by combining the section of barometric and other curves at Hamburg with the records of other observatories and the synoptic charts of Germany, at 8 p.m. the same evening, in the manner that we explained in our chapter on Meteograms, has built up the beautiful diagram of this storm, which we reproduce, with a few trifling modifications, so as to assimilate it with our other illustrations, in Fig. 51. There the isobars are given both in millimetres and their approximate equivalents in inches: *t* marks the line of the trough; the shaded line *r* the position and dimensions of the rain-stripe; the long dotted arrow the direction in which the whole system was being propagated; and the small solid arrows the direction and force of the wind across any section. The whole is evidently, in the main, one of those V's in which the rain begins just after the passage of the trough; but the curious projection upwards of the isobars just under the rain-stripe is unlike anything we have seen before; and the strong west wind, with four feathers on the arrow, is quite unconformable with the isobars according to our usual experience. From whence comes all this? First, for the upward projection of the isobars. All the barographs showed,

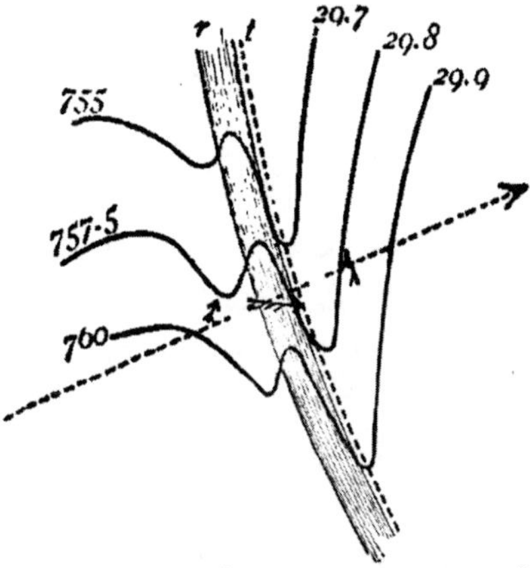

Fig. 51.—Line-thunderstorm. *t*, Trough of V-depression; *r*, rain stripe.

about seven minutes after the passage of the trough, a sudden rise exactly similar to those marked *b* in Fig. 48 just as the heavy rain began, and this rise, as usual, was quite distinct from the general increase of pressure due to the rear of the V. If we were to superimpose a long, narrow, isolated ridge of high-pressure on the rear of a V, we should get an inverted V, or wedge, exactly like that which we find in the diagram under the rain-stripe; but we must not treat this like the ordinary wedge-shaped isobars which we have before described. The form is the same, but the cause is different.

Then for the wind-sequence. We find in front of the V a light south-east wind; then, just before and at the commencement of the rain, a very violent squall (called "boe" in Germany) from the west; and, finally, a light south-west wind in rear of the whole disturbance. The westerly squall is nearly perpendicular to the isobars and to the lie of the rain-stripe; but, as the isobars here are not the lines of general atmospheric circulation, but are partly due to purely local causes, we need not be surprised that Buy Ballot's law does not hold.

Temperature was, as usual, very high in front; very low for the season in rear of the trough. If we could have drawn the isotherms, we should have found them running almost north and south, nearly parallel to the rain-stripe. It is impossible to determine at present how much of this cold is due to the mechanical transport of cold air with the heavy rain, and how much to the general descent of cold air in rear of the V as a whole.

The rain-stripe, we see, was in this instance a long narrow band, moving broadside on, and manifestly connected with the trough of a V-depression.

We will now give an example of line-thunderstorms which are not associated with the trough of either a V or a cyclone, though they also move broadside on, nearly perpendicular to the depression with which they are in some way associated. In Fig. 52 we give synoptic charts for France at 9 a.m. and 9 p.m. (Paris time) on August 21, 1879. The full lines are isobars, a few arrows show the general direction of the wind, and the one dotted line in each marks the mean position of the thunderstorms which were raging at that moment. In France the mean of the time between the first and last thunder is taken to give the position of the storm at a given hour. For instance, if the first thunder was heard in a place at 8 a.m. and the last at 10 a.m., then 9 a.m. would be marked as the hour at which the storm passed the station. This method is manifestly inferior to that of noting the times of first and last thunder, and then plotting the shape of the storm on a chart. The barometric changes during the day were

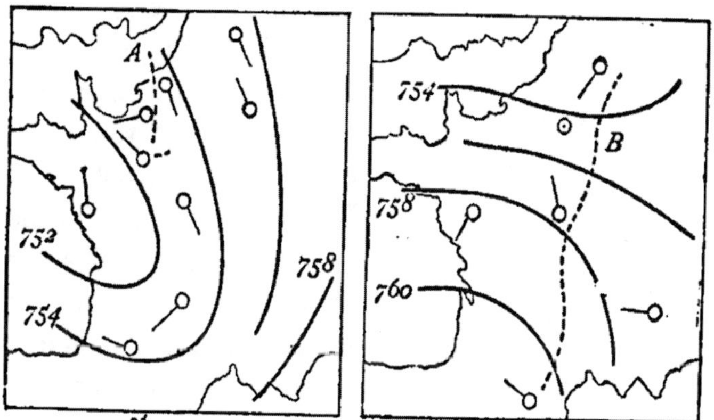

Aug. 21st 1879. 9 a.m. *Aug. 21st 1879. 9 p.m.*

FIG. 52.—Thunderstorms in France.

really much more complicated than might appear from an inspection of the maps over the limited area of France. The large secondary cyclone whose centre lay over the west of France in the morning, appears to have crossed that country in a north-easterly direction, and to have merged in a complicated manner with a larger depression which lay to the north-west of Ireland in the early morning. By 9 p.m., however, the whole of the south-west of France was covered by an anticyclone, whose origin we cannot trace.

In Fig. 53 we give a diagram of the positions at every alternate hour of two sets of thunderstorms that traversed

France during that day, which are marked A and B respectively. Such lines are called "isobrontons," or lines of equal development of thunder. Those in the figure may be taken as typical of the march of this class of thunderstorm all over Europe—a long narrow line, advancing broadside on towards the east or north-east, pretty nearly independent of the shape of isobars with which it is associated.

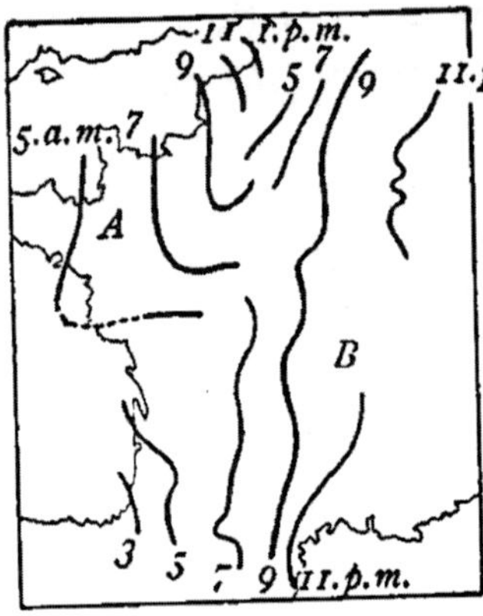

FIG. 53.—Track of thunder-storms.

The first storm, marked A, struck the coasts of the Bay of Biscay at five o'clock in the morning. The shape of the front was bent, the ends being most in advance: this is very often the case in France. The position of the front of this storm is given for every two hours' interval until 1 p.m., when the disturbance appears to have died out near Calais. The relation which the position of the thunder bore to the secondary cyclone may best be seen by reference to the preceding chart (Fig. 52). There the dotted line shows that the front of the storm A lay to the north-east of the centre of the cyclone; and, as far as the small number of wind-arrows allow us to judge of the nature of the disturbance, there seems to be a small local deflection of the wind in rear of the storm. The wind in rear should have been from south-east with the shape of isobars there represented, whereas the chart shows that in some places the direction was from the west and north-west.

The second series of thunderstorms, marked B, commenced at Biarritz at three o'clock in the afternoon, and moved irregularly in a more westerly direction than the morning storm. The storm, in its course during the day, seems to have increased enormously, for at 9 p.m. we find the front reaching nearly from Brussels to Perpignan in the Pyrenees, a distance of six hundred miles. This will perhaps enable us to realize the disastrous character of hail and thunderstorms in France. Here we have a line of destruction six hundred miles long, and from ten to twenty miles broad, sweeping like a curtain across the country at a rate of about thirty miles an hour, wrecking in a few minutes vineyards which are worth many thousands of pounds, and destroying at the last moment the husbandman's labour for the whole year.

But now we come to one of the most puzzling points connected with thunderstorms. If we look at Fig. 52, where we have marked by a dotted line the position of the storm-front B at 9 p.m., it is very difficult to see any connection between the shape of the isobars and the position of the thunder. So far as we can see, there is no trace of the trough either of a cyclone or of a V-depression, and all the general indications would have been for improving weather after the passage of the secondary cyclone. Whether more numerous observations at stations nearer to one another would have shown the presence of secondaries, we cannot say; but this is by no means an isolated instance in France, and similar cases occur in other countries. For the present, therefore, the explanation of the nature of this class of thunderstorms must await future research; all that we can do here is to note them as an apparent exception to the general course of weather which we have already explained. We must also specially note them as cases where rain falls with a steady rising barometer; and also understand that, although a forecaster could not have pointed out in the morning exactly when or where thunderstorms would strike, he could have said with certainty that storms would occur during the day in many parts of France. Secondaries can never form in summer without some electrical disturbance.

The details of rain and cloud in line-squalls and thunderstorms are extremely interesting, and for them we are chiefly indebted to the careful researches of Dr. Koppen. The approach of a thunder-squall usually announces itself by the rapid crowding up of heavy clouds. We see a dark, black border, often looking like a long roll or wreath, and beyond this a peculiar light grey uniform sky. The dark, low band passes overhead, and heavy rain commences as the light grey cloud comes on. The first burst of rain is usually the heaviest, and after a longer or shorter period the rain usually clears gradually off. This is the rain with which the sudden rise of the barometer is observed. The wind, which has fallen very light from south-east or south as the clouds begin to bank up, comes in a violent squall from the west, about the time the dark wreath passes overhead, and falls again shortly after the commencement of the heavy rain. A good illustration of a very pronounced cloud-wreath will be found in Fig. 56 in the next chapter under "Pamperos." We give an ideal diagrammatic section of such a squall in Fig. 54. We may suppose that from general causes, such as the trough of a V-depression, a cold westerly current meets a warmer one from the south-east or south. The latter rises, as shown by the small arrows, and curls over where the black wreath (*w*) of cloud is found, and then the commingling of the two currents forms a gigantic dark vault (*v*) of cloud, from which heavy rain (*r*) pours down. The light

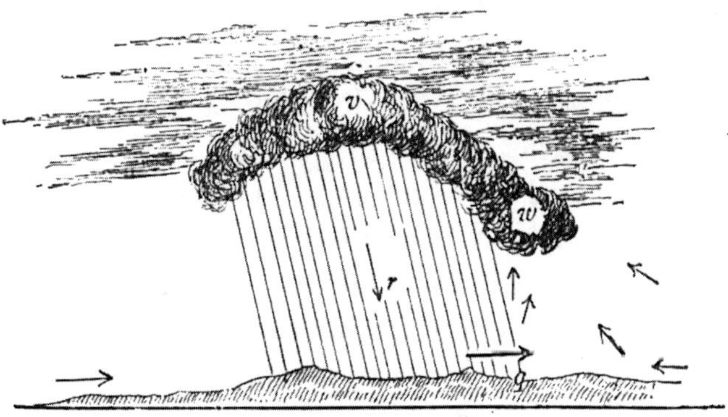

FIG. 54.—General circulation and cloud-vault of line-squall.

grey cloud which an observer sees behind the black wreath is really a peep into the rain falling from this great vault. The big drops of rain bring down mechanically with them a vast amount of cold air, which rushes straight out in front of the squall—it has no time to pick up anything from the earth's rotation—and produces the squall *q*, marked by a long arrow.

Our section, then, of a squall is that of a vertical whirl, the whole system perhaps not one mile high by one and a half mile across, while the length of the front of the storm may be two hundred miles; and our picture of the whole must be a long, nearly straight horizontal axis, moving broadside on, round which the wind whirls vertically in a direction opposite to that of the watch-hands. It is always an episode, as it were, in the history of some general form of atmospheric circulation. We can, perhaps, see how a broad-fronted west current can meet a southerly stream of air in the trough of a V or cyclone; but we are unable at present to form any conception of a straight-fronted current advancing across the curved isobars of an anticyclone, as in Fig. 52 for French thunderstorms.

Line-squalls and thunderstorms of a different type are very common in the tropics. The author has observed a very striking instance at the junction of the south-east trade and north-west monsoon in the Indian Ocean. There was no doldrum, but the two currents met along a line whose position was marked by a long dark, black cloud, with heavy rain and squall.

In like manner, the daily thunderstorm which occurs in so many countries at the time when the sea-breeze comes in, charging, as it were, the prevailing wind over the land, is due to a long vertical whirl where the two currents meet, and the whole length can sometimes be watched gradually advancing inland from the coast. The so-called north-westers at Calcutta, during the hot season, belong to this last type.

Thunderstorms with Secondaries.

We must now just mention a class of thunderstorms which are more complicated than a simple squall, and yet differ in many ways from line-thunderstorms. They are associated with secondary cyclones, and are much commoner in England than line-thunderstorms, but none have been tracked over a sufficiently long area to allow us to say anything about their shape or motion. All we know is, that as surely as we see a secondary on the charts in summer, so certainly will thunderstorms occur during the day, though we cannot say in what portion of the small depression.

The general features of this kind of disturbance will be best understood by reference to Fig. 55, where we give a typical example of that distribution of pressure which is associated with a summer thunderstorm in Great Britain. The isobars, wind-arrows, and weather-symbols give the synoptic conditions of North-western Europe at 8 a.m. July 3, 1883. The broad features of pressure-distribution are found in an anticyclone over Scandinavia, in the fragment of a large depression to the west of Ireland, and in a complicated mass of secondaries over Great Britain and the north of France. The general direction of the wind is from the south, but both the bends in the isobars, which mark out the position of the secondaries over Central England and the north of France, are associated with a considerable deflection of

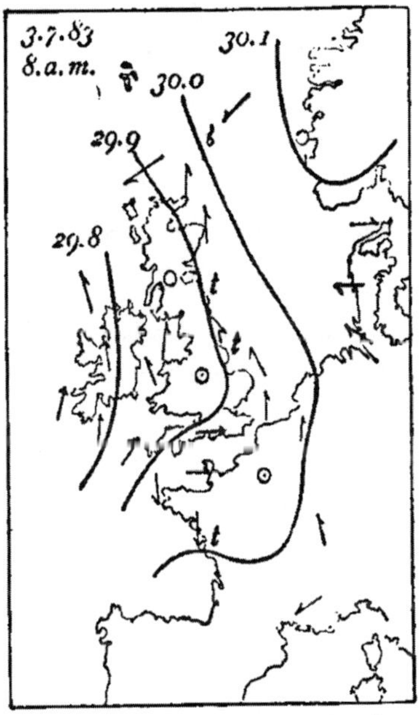

FIG. 55.—Conditions of thunderstorms.

the wind. In both instances we see a partial circulation of the wind round a central spot of calm; the latter stations are marked by the symbol of a circle with a dot in the centre. At the moment when the chart was constructed, three thunderstorms were in progress: two on the east coast of England—at Shields and Spurn Head—with one in France, at L'Orient. All these stations are marked with the symbol *t*.

The special features of this class of thunderstorm are the calm sultry weather with which they are associated, so different from the squall of a line-thunderstorm, and the limited rotation of the surface-wind during the progress of the storm. Another very remarkable feature is that this surface circling of the wind extends only a very short distance upwards, and whenever a glimpse can be caught of the drift of the upper clouds, they are found to move in the same direction throughout the whole period of the disturbance.

This is the familiar class of thunderstorm which we associate with sultry weather, and with the thunder coming against the wind. As the secondary approaches any station, the wind draws more or less in towards the centre, and recovers its former direction after the depression has passed. In a case like that figured in the diagram, the motion of the storm as a whole would be towards the north-east, and the rotation of the wind at any place would depend on which side of the centre the station lay.

Squalls, troughs, and secondaries do not by any means exhaust all the conditions under which thunder and lightning may be developed, even within the limits of Europe, but any attempt to examine into these other causes would be beyond the scope of this work. We have, therefore, only indicated the three principal sources of electrical discharge, which, however, will account for more than eighty per cent. of European thunderstorms. The great thing is to realize that there are several different kinds of thunderstorms, each of which has certain distinctive features.

General Remarks.

We may, perhaps, conclude this short notice of thunderstorms with a few general remarks on the subject. One of the first things which must strike everybody is, that even in the temperate zone some countries are far more ravaged by thunderstorms than others. For instance, France suffers more than any other part of Europe, and England the least. We may probably find at least two causes which modify the development of thunderstorms. In the first place, the geographical position of the country relative to the great seasonal areas of high and low pressure. From this point of view we can readily see that France is far more exposed to the influence of small secondaries, which come in from the Atlantic, and which die out before they reach Central Europe, than any other portion of that continent. If we look at the large charts which we give in our chapter on Weather Types, we can easily understand that whenever a cyclone leaves the Atlantic anticyclone to go towards the north-east, there must be left a somewhat irregular col of pressure over France, and this we know is most conducive to the formation of thunderstorms.

The second cause which may modify the formation of storms is the amount of vapour in the air. We know by experiment that the discharge of frictional electricity is very much modified by the hygrometric condition of the atmosphere. In England, if we turn a machine on a damp day, the electricity will escape as fast as it is made, and no discharge can be obtained; on a dry day, on the contrary, sparks can easily be procured. From this analogy we can easily conceive that the same atmospheric disturbance which causes a violent thunderstorm in the dry climate of France, would either discharge its electricity noiselessly, or at all events occasion but a very feeble storm, in the damper atmosphere of Great Britain.

No doubt, insulated damp air is just as good a non-conductor as dry air; and the silent discharge of a frictional machine in damp weather is due to the condensation of a thin layer of water on the surface of the insulating supports. But in the atmosphere we cannot conceive absolute insulation. The air is full of ice or water-dust, and we know that somehow or other electrical discharge is easily propagated from one cloud to another.

We cannot say absolutely, in our present state of knowledge, how far electricity is only a secondary product of atmospheric disturbance, or whether it ever plays a primary part in making a storm. As far as we can see, however, the evidence is entirely in favour of the idea that electricity is only a secondary phenomenon. We cannot suppose that an abnormal amount of electricity can ever be developed without some definite cause, and it is also an obvious fact, that in thundery weather there are often a great many showers without thunder, which only differ from those with thunder and lightning in violence or intensity.

On the other hand, it is almost certain that the discharge of electricity is in some manner associated with the formation of hail and very large rain-drops; but at present no complete explanation can be given of the relation between these two phenomena.

Non-Isobaric Rains.

We have now to deal with the most unsatisfactory branch of modern meteorology—the nature of those falls of rain that are not associated with any definite shape of isobars, and which are therefore called "non-isobaric" rains. We have already described the very remarkable case of line-thunderstorms whose position cannot be detected by any inspection of isobars; and numerous other cases of a less striking nature are of constant occurrence. The importance of this classification of rains in any comprehensive treatise on meteorology may be judged from the fact that in Great Britain, though the bulk of winter rain is cyclonic, a great deal of summer rainfall is non-isobaric; in Continental Europe a still larger proportion is of the latter character; so are most tropical rains, except the downpour of hurricanes; while the whole of the heavy rain on the equator, and all that alls in the doldrums, is also absolutely non-isobaric.

The South-west Monsoon.

But by far the most striking non-isobaric rain in the world is the burst of the south-west monsoon in the

Indian Ocean. Let us consider the course of the seasons in Ceylon, and correlate them with changes of pressure over India. In the month of February we find a very shallow stationary depression—not a cyclone—over Lower Bengal, a belt of high pressure stretching across the Bay of Bengal from Madras to Rangoon, and a general diminution of pressure from that belt to the equator. From this we might reasonably expect what we find—light south-west wind over Lower Bengal, variable breezes over Madras, and a light north-east monsoon across Ceylon; but it is not so obvious why the south-west wind should be so fine and dry as it is. The low pressure over Bengal gets gradually more pronounced, and spreads with its accompanying south-west wind slowly southwards, till Ceylon is embraced within its sphere. These conditions are most pronounced towards the end of May, and we get the dry, nearly cloudless, hot season of India and Ceylon with a light south-west wind. Then the sky begins to cloud over, and suddenly rain bursts in a series of terrific thunderstorms, and the bad wet weather continues for two or three months. The rain begins in Ceylon, and then works slowly up the west coasts of India and Burmah —omitting Madras—till Calcutta and Lower Bengal are reached, three or four weeks later than Colombo. Then we are met by the strange fact that this, the most striking weather-change in the whole year, is associated by no change in the shape of the isobars. We give in our chapter on Weather Types two diagrams (Figs. 82 and 83) of Indian isobars just after the monsoon has burst over Bombay and Calcutta. The only difference in the isobars then and a fortnight previously, when the hot dry season prevailed, is that the level of pressure is a little lower, that the position of lowest barometer has moved a little higher up the Ganges, and that the distortion of the isobars by secondaries is more pronounced. When the monsoon is fairly established, we can, no doubt, see certain slight fluctuations in the shape and intensity of the isobars which accompany what is called "a break in the rains," and sometimes exceptionally heavy rain falls during the passage of a small cyclone from the Bay of Bengal up country; but we cannot find any change in the isobars to account for the sudden change of weather which is called in common parlance, "the burst of the monsoon." The quality of the rain, if nothing else, distinguishes the monsoon from cyclonic precipitation. The rain in front of a Bengal cyclone seems to grow out of the air, while that of the monsoon falls in thunderstorms and from heavy cumuloform clouds. The only rational suggestion which has been made to account for this burst of rain, would look to a sudden inrush of damp air from the region of the doldrums as the source of the change in weather, but not of the direction of the wind, or of the shape of the isobars; for the burst is apparently almost coincident with the disappearance of the belt of high pressure to the south of the Bay of Bengal.

No satisfactory clue has, however, yet been discovered either to the cause, or still less to the quantity, of non-isobaric rainfalls. They are the bugbear of every European forecaster, though in Japan, curiously enough, they find rain easier to announce than the direction of the wind. Mr. Finley, of the United States Signal Office, has made some very interesting studies on local rains that do not show on the isobaric charts. He takes a map of the United States, and puts in all the wind-arrows without any isobars. Very often he finds some large areas swept by a generally southerly wind, and others by a generally northerly wind, and he draws lines to mark out the tracts of country where these currents meet, and where they diverge. Then he finds that there are always local rains over the first areas, and rarely any over the latter. This would undoubtedly point to local vertical whirls between the meeting currents as the source of rain.

Whether this is universally the case, or whether the conditions of all rains could be analyzed into small V's or secondaries if the isobars were constructed from stations sufficiently close together, we cannot at present say. The important thing is not to mix up all kinds of rain together when we want to discuss general meteorological problems.

CHAPTER IX.

PAMPEROS, WHIRLWINDS, AND TORNADOES.

WE will now describe two remarkable kinds of storms which occur in La Plata and in the United States respectively.

PAMPEROS.

The word "pampero" is, unfortunately, used in a very vague manner in the Argentine Republic and neighbouring states. Every south-west wind which blows from off the pampas is sometimes called a pampero; and there is a still further confusion caused by calling certain dry dust-storms *pamperos sucios*, or dry pamperos. The true pampero may be described as a south-west wind, ushered in by a sudden short squall, usually accompanied by rain and thunder, with a very peculiar form of cloud-wreath. We will describe these as given by D. Christison in the *Proceedings of the Scottish Meteorological Society*, No. lx. p. 330, and then we shall have no difficulty in recognizing a line-squall as the source of the pampero.

The barometer always falls pretty steadily for from

two to four days before the pampero, and always rises for some days after the squall. There are not enough barometric observations to allow of any generalizations as to the precise position of the squall relative to the trough of the general depression, but in two recorded cases the mercury began to rise some hours before the storm burst.

Temperature is always very high before the squall, and then the sudden change of wind sends the thermometer rapidly down, sometimes as much as 33° in six hours.

Thunder accompanies about three out of four pamperos; but more or less rain always falls, except in the rarest cases.

The wind before this class of pampero almost invariably blows moderately or gently for some days from easterly points, and then with a sudden burst the south-west wind comes down with its full strength, and, after blowing thus from ten to thirty minutes, either ceases entirely or continues with diminished force for a certain number of hours. In all cases but one the upper wind-currents have been seen to come from the north-west both before, during, and after the pampero.

The general appearance of a pampero will be best understood by a description of an actual squall. "In the early morning of a day in November, the wind blew rather strongly from the north-east. The sky was cloudy, but not overcast, save in the south-west horizon. The clouds were moving very slowly from the west, or a little south of it, throwing out long streamers eastwards. About 8 a.m. the threatening masses in the south-west had advanced near enough to show that at their head marched two dense and perfectly regular battalions of cloud, one behind the other, in close contact, yet not intermingling, and completely distinguished by their striking difference of colour, the first being of a uniform leaden grey, while the second was as black as the smoke of a steamer. On arriving overhead, it was seen that the front, although slightly sinuous, was perfectly straight in its general direction, and that the bands were of uniform breadth. As they rushed at a great speed under the other clouds without uniting with them, preserving their own formation unbroken, their force seemed irresistible, as if they were formed of some solid material rather than vapour. The length of these wonderful clouds could not be conjectured, as they disappeared beneath the horizon at both ends, but probably at least fifty miles of them must have been visible, as the 'Cerro' commands a view of twenty miles of country. Their breadth was not great, as they only took a few minutes to pass overhead, and appeared to diminish from the effects of perspective to mere lines on the horizon. At the instant when the first band arrived, the wind—which was still blowing, and something more than gently, from the north-east—went round by north to south-west; at the same time a strong, cold blast fell from the leaden cloud, and continued to blow till both bands had passed. From neither of them, however, came lightning or rain, but, filling up the sky in rear of the regular army, followed a confused rabble of clouds, with a constant rumbling of thunder, and from which evidently rain was falling. It was not, however, till fifteen minutes after the passage of the two regular bands that rain fell where the observations were taken. The storm, passing on, obscured the whole sky, wind, rain, and thunder continuing for some hours, but only to a moderate degree." The diagram (Fig. 56), taken from a sketch made at the time, represents the northerly half of the storm-clouds while still at some distance from the spectator, and advancing from a westerly direction.

From all this it is manifest that the changes of wind, the rapid alternations of temperature, and the typical cloud-wreaths are identical in character with the class of

FIG. 56.—Cloud-wreath in pampero.

disturbance we have described in the previous chapter under the heading of Line Squalls and Line Thunderstorms. We must remember, however, that being south of the equator, north-east, south-west, and north-west winds are equivalent to those from south-east, north-west, and south-west in the northern hemisphere.

Whirlwinds.

A whirlwind may be described as a mass of air whose height is enormously greater than its width, rotating rapidly round a more or less vertical axis. A moderate whirlwind may be two hundred feet high, and not above ten feet in diameter. The dimensions, however, are very variable, for a whirlwind may vary in intensity from a harmless eddy in a dusty road to the destructive tornado of the United States. As the latter are the most terrific manifestation of weather in the whole range of meteorology, we shall devote a few pages to their consideration.

Tornadoes.

A tornado is simply a whirlwind of exceptional violence; if it were to encounter a lake or the sea, it would be called a waterspout. Its most characteristic feature is a funnel, or spout, which is the visible manifestation of a cylinder of air that is revolving rapidly round a nearly vertical axis. This spout is propagated throughout the northern temperate zone in a north-easterly direction at a rate of about thirty miles an hour, and tears everything to pieces along its narrow path.

The diameter of the actual spout often does not exceed a few yards, and the total area of destructive wind is rarely more than three or four hundred yards across. The height of the spout is that of the lowest layer of clouds, which are then never high; and, as in thunderstorms, the upper currents are unaffected by the violent commotion below.

The spout as a whole has four distinct motions:

1. A motion of translation generally towards the north-east at a variable rate, but which may be taken to average thirty miles an hour.

2. A complex gyration. The horizontal portion of this rotation is always in a direction opposite to that of the hands of a watch—that is to say, in the same manner as an ordinary cyclone. But in addition to this there is a violent upward current in the centre of the cylinder of vapour or dust which constitutes the spout, and sometimes small clouds seem to dart down the outer sides of the funnel whenever these float in close proximity. There are, however, no authentic instances of any object being thrown to the ground by the individual effort of a downward current. The slight downward motion of a few small clouds is probably only a slight eddying of a violent uprush.

3. A swaying motion to and fro like a dangling whip, or an elephant's trunk, though the general direction of the spout is always vertical.

4. A rising and falling motion, that is to say, that sometimes the end of the funnel rises from the surface of the ground and then descends again, and so on. Owing to this rise and fall, the general appearance of the tornado changes a good deal. When the bottom of the spout is some distance above the ground, the whole is somewhat pointed, and does comparatively little harm as it passes over any place. As the spout descends, a commotion commences on the surface of the ground. This latter gradually rises so as to meet the descending part of the spout, and then the whole takes the shape of an hour-

glass, as in Fig. 57. This is the most dangerous and destructive form, because the ground gets the whole force of the tornado.

The general appearance of the cloud over a tornado or whirlwind is always described as peculiarly smoky, or

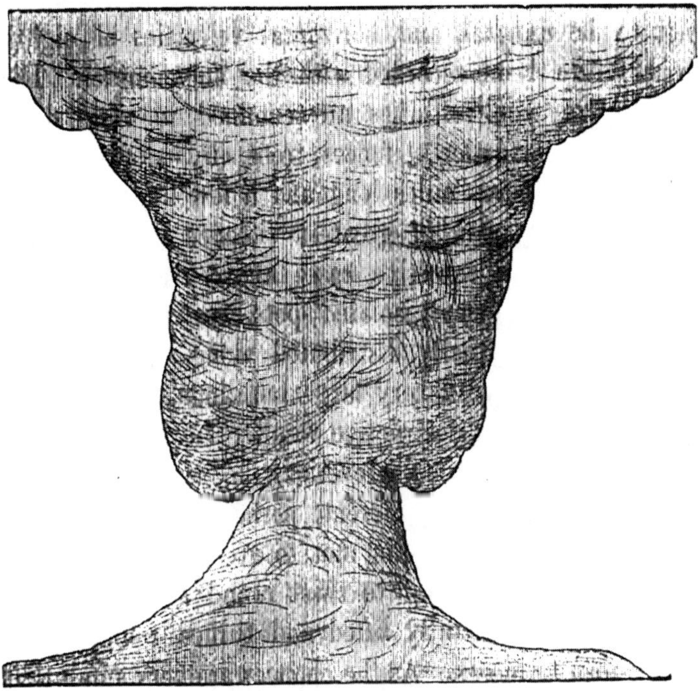

FIG. 57.—Tornado-cloud.

like the fumes of a burning haystack. The tornado is also never an isolated phenomenon; it is always associated with rain and electrical disturbance.

The destructive effects of the tornado are very curious, from the sharp and narrow belt to which the injury is confined. It appears that in the passage of some tornadoes wind-pressures of various amounts, from eighteen to a hundred and twelve pounds per square foot, have been demonstrated by destruction of bridges, brick buildings, etc. The upward pressures are sometimes as great as the horizontal, and even greater. Downward pressures or movements of wind have not been clearly proved. Upward velocities of 135 miles per hour seem not to be unusual, and horizontal velocities of eighty miles have been recorded with the anemometer. The destructive wind-velocities are confined to very small areas. A destruction of fences, trees, etc., is often visible over a path many miles long and a few hundred yards wide, but the path of greatest violence is very much narrower. The excessive cases above referred to are observed only in small isolated spots, less than a hundred feet square, unequally distributed along the middle of the track. Thus, in very large buildings, only a small part is subject to destructive winds. In different parts of this area of *maximum* severity, the winds are simultaneously blowing in different, perhaps opposite, directions, the resultant tending not to overturn or carry off or crush in, but rather to twist round a vertical axis. Buildings are generally lifted and turned round before being torn to pieces. As the chances are very small that a building will be exposed to the violent twisting action, it is evidently the average velocity of rectilinear winds within the path of moderate destruction that it is most necessary to provide against in ordinary structures. These winds may attain a velocity of eighty miles an hour over an area of a thousand feet broad, and generally blow from the south-west; the next in frequency blow from the north-west. The time during which an object is exposed to the more destructive winds varies from six to sixty seconds. An exposed building experiences but one stroke, like the blow of a hammer, and the destruction is done. Hence, in a suspension-bridge, chimney, or other structure liable to be set into destructive rhythmic vibrations, the *maximum* winds do not produce such vibrations. The duration of the heavy south-west or north-west winds over the area of moderate destruction is rarely over two minutes. The motion of translation of the central spout of a tornado, in which there is a strong vertical current, is, on an average, at the rate of thirty miles an hour.

Tornadoes mostly occur on sultry days and either in the south-east or right front of cyclones, or in front of the trough of V-depressions.

The relative frequency of tornadoes is, in order of decreasing frequency, June, July, April, May, . . . January. In the geographical distribution of 247 tornadoes from 1794 to 1878, the largest figures are obtained from New York (24), Indiana (20), Illinois (20), Ohio, and Georgia (16 each), etc.; but the records are fragmentary, and now Kansas is the most tornado-stricken state. The largest number of tornadoes apparently occur between 5 p.m. and 6 p.m.; the next between 4 p.m. and 5 p.m.

The following example will illustrate all the principal features of tornadoes. It is taken from one of the reports of the United States Signal Office, and the author is indebted to the chief signal-officer for supplying him with the materials. The most important part of the work has been done by Mr. Finley of that office, who for many years has devoted his special attention to the subject.

In Fig. 58 we give a synoptic chart of the meteorological conditions of a portion of the United States on May 5, 1879, at 4.35 p.m., Washington time. The course of the great rivers Mississippi, Missouri, and Ohio, together with the outline of Lake Michigan, are clearly marked

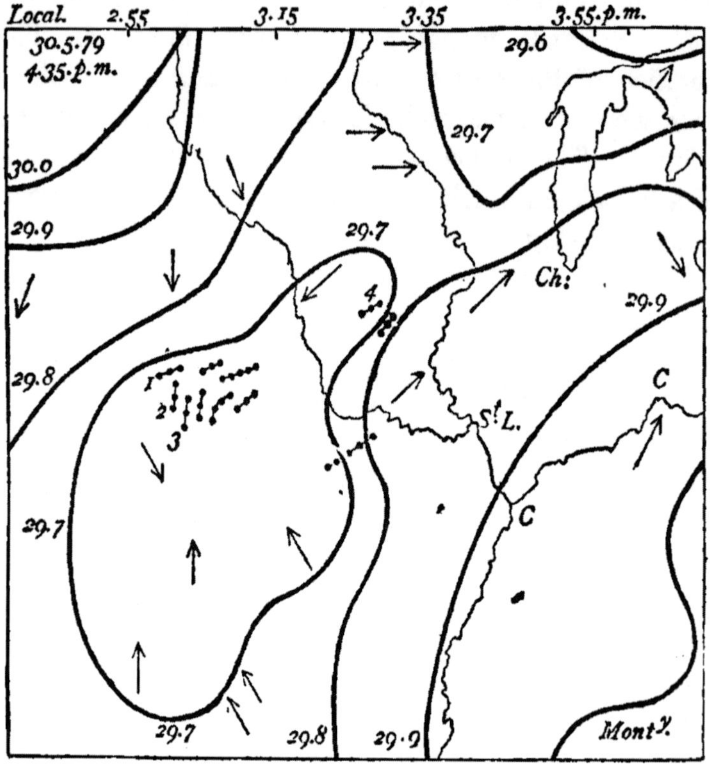

FIG. 58.—Conditions and paths of tornadoes.

by thin lines; and the position of the cities of Chicago, St. Louis, Cincinnati, Cairo, and Montgomery will be readily recognized by their initial letters; so that there is no need to indicate the boundaries of any state. The isobars are shown by bold black lines, and the arrows fly with the wind at the different stations. No less than eleven tornadoes ravaged the Western States during that day; the course of each of these is shown on the map by a line of small circles, joined together, and those numbered 1, 2, 3, and 4 respectively were apparently actually in progress at the moment to which the chart refers. Local time is indicated by figures on the top of the diagram.

The broad features of pressure-distribution are sufficiently simple. One area of high pressure lies over Manitoba; another over the southern states; a col lies between them. On the previous day this col had been filled by a V, which by this morning had partially developed into the secondary cyclone which now lies over the Western States. A portion of the original V is still seen just to the west of Lake Michigan. The trough of the V is undoubtedly connected with the trough of the secondary, but the latter appears to be made up of several subsidiary depressions; the general direction of the wind is typical of such isobars. In front of the V and secondary the general sweep of the wind is from the south—a little more south-east in some places, and a little more south-west in others.

In rear of the trough, the general direction of the wind is northerly—a little more north-east in some places, and a little more north-west in others. The formation of tornadoes appears to have been associated everywhere with the secondary, but at the moment for which the chart is constructed, the tornado marked 4 seems to have been produced by a different disturbance from that which caused numbers 1, 2, and 3. If we had more barometric observations, we should most probably find that the projection of the isobar of 29·7 ins. which surrounds number 4 tornado, was really a separate subsidiary depression. It should also be noticed that all the eleven tornadoes moved in the same general direction, which was practically identical with that of the whole system of depressions.

The general character of all tornadoes is so similar that the description of one will do for all. We shall therefore give some of the description furnished by an eye-witness to the United States Signal Office, of the tornado marked 3 in the chart (Fig. 58), which is described in the reports as the "Delphos" tornado.

"On Friday morning, May 30, 1879, the weather was very pleasant, but warm, with the wind from the south-east, from which direction it had blown for several days. The ground was very dry, and no rain had fallen for a number of weeks. About 2 p.m. threatening clouds appeared very suddenly in the west (against the wind), attended in a few minutes by light rain, the wind still in the south-east. It stopped in about five minutes, and then commenced again, wind still the same, accompanied by hail, which was thick and small at first, but rapidly grew less in quantity and larger in size, some stones measuring three and a half inches in diameter, and one was found weighing one-fourth of a pound. This last precipitation continued for about thirty minutes, after which a cloud in the shape of a water-spout was seen forming in the south-west, and moving rapidly forward to the north-east. The cloud from which the funnel depended, seen at a distance of eight miles, appeared to be in terrible commotion; in fact, while the hail was falling, a sort of tumbling in the clouds was noticed as they came up from the north-west and south-west, and about where they appeared to meet was the point from which the funnel was seen to descend. There was but one funnel at first, which was soon accompanied by several smaller ones, dangling down from the overhanging clouds like whip-lashes, and for some minutes they were appearing and disappearing like fairies at a play. Finally one of them seemed to expand and extend downwards more steadily than the others, resulting at length in what appeared to be their complete absorption. This funnel-shaped cloud now moved onward, growing in power and size, whirling rapidly from right to left, rising and descending, and swaying from side to side. When within a distance of three or four miles, its terrible roar could be heard, striking terror into the hearts of the bravest." The eye-witness judged that the funnel itself would reach a height of about five hundred feet from the ground. As the storm crossed a river, a cone-shaped mass came up from the earth to meet it, carrying mud, *débris*, and a large volume of water (Fig. 57). The cloud then passed the observer's house very near to 4 p.m. The progressive velocity at the time was considered to be about thirty miles per hour, although at Delphos, three and a half miles distant, it had slackened down to near twenty miles. A few minutes previous to and during the passage of the funnel, the air was very oppressive; but ten minutes after, the wind was so cold from the north-west that it became necessary to wear an overcoat when outside. As the tornado struck the house, another member of the family says, "We think it is coming near us. We can now see its fury. Shall we leave the house? No; for we are not certain on which side it will pass. We are apparently as safe here as elsewhere. The windows are nailed fast. Three of us lean against the door which is nearest the storm; the rest go into the cellar. It is about 4 p.m. A moment of breathless suspense, and the storm strikes us. The timbers creak, the sides of the house sway in and out; surely they cannot outlast it. We hear no well-defined roar now, for on the outside boards and other *débris* are fiercely crashing. All is dark within. In about fifteen minutes the storm is over. We leave the house. The centre of the storm has passed to the west of us, and we can see its dark form moving away in a north-east direction."

The actual diameter of this storm appears to have been only forty-three yards. On the right of the track, destructive winds extended to a further distance of from one to two miles, sensible deflected winds for another mile and a half, beyond which only the usual wind of the day was experienced. On the left or northern side of the tornado path, the damage did not extend quite so far, for the width of the belt of destructive winds was not more than twenty-eight yards across, and that of sensibly deflected winds one mile and a quarter.

As a specimen of the damage done, a large two-horse sulky plough, weighing about seven hundred pounds, was carried a distance of twenty yards, breaking off one of the iron wheels attached to an iron axle one and three-quarter inches in diameter. A woman was carried to the north-west two hundred yards, lodged against a barbed-wire fence, and instantly killed. Her clothing was entirely stripped from her body, which was found covered with black mud, and her hair matted with it. A cat was found half a mile to the north-west of the house, in which she had been seen just before the storm, with every bone broken. Chickens were stripped of their feathers, and one was found three miles to the north-west.

A few miles further on, another eye-witness says, "The dark, inky, funnel-shaped cloud rapidly descended to the earth, which reaching, it destroyed everything within its grasp. Everything was taken up and carried round and round in the mighty whirl of the terrible monster. The surrounding clouds seemed to roll and tumble towards the vortex.

"The funnel, now extending from the earth upwards to a great height, was black as ink, excepting the cloud near the top, which resembled smoke of a light colour. Immediately after passing the town, there came a wave of hot air, like the wind blowing from a burning building. It lasted but a short time. Following this peculiar feature, there came a stiff gale from the north-west, cold and bleak, so much so that during the night frost occurred, and water in some low places was frozen."

Relation of Whirlwinds to Cyclones.

Before concluding this chapter, we may make a few remarks on a very interesting question which here presents itself. Commencing with a whirlwind only two or three feet across, we find every gradation of size till we come to the destructive tornado. From the small secondary which deflects the wind in connection with a thunderstorm, there seems to be every gradation of size into the secondary which is so large that we can hardly say whether it should not be called a primary cyclone.

In both the whirlwind and cyclone series we have certain common features—a horizontal rotation, and more or less uptake near the centre of the gyration. But is there any intermediate series between the whirlwinds and the cyclones, or can the former ever develop into the latter? We believe not, though the opposite opinion has often been propounded.

In the first place, we are unable to find any connecting link between the two types of rotation. Under certain conditions, the wind seems to have a tendency to form little eddies, which under favourable circumstances may grow into complete cylinders of rotating dust. In our chapter on Prognostics we have shown that in England whirling dust is a well-known precursor of showers of rain, but not of the true cyclone-rain. In other countries, such as the Punjab and on the Isthmus of Suez, regular whirlwinds are of daily occurrence at certain seasons of the year; but these never by any chance grow into even the smallest secondary. We have just seen that the terrific tornadoes of the Western States of the American Union are merely an episode in the conflict of opposing currents near the trough of larger depressions, but the whirlwinds never give rise to any larger disturbance. In every one of the eleven tornadoes which occurred on the day we have just described in some detail, it was found that rain and hail invariably preceded the tornado-cloud from ten to thirty minutes, and that the tornado was only, as it were, a local accident in a very large disturbance.

In our chapter on Weather-types, we shall give abundant illustrations of the manner in which both primary and secondary cyclones are formed without the presence of two such opposing currents as are found in front and rear of the trough of a V-depression, and we shall see how the two kinds of cyclones may either develop or degrade from one into the other.

We have also already seen the very small circulation of the wind which accompanies some kinds of thunderstorms, but in no case do we find any transitional link between the whirlwind and cyclone types of rotatory motion.

At the same time, we may note that the inner core and very deep central depression of a tropical hurricane approximates more nearly to the tornado type than the cyclones of temperate regions; but the absence of transitional forms seems conclusive against the identity of tornadoes and cyclones. In both the destructive fury is out of all keeping with any forces that we are acquainted with; and their true nature remains to be discovered by future research.

CHAPTER X.

LOCAL VARIATION OF WEATHER.

Nature and Principles.

The object of this chapter is to explain what is known as the local variation of weather. This term groups conveniently a large class of dependent phenomena, which owe their origin to the influence of local obstacles or peculiarities on the development of weather. We know that in the same country some places are much colder or wetter than others; that some are more exposed to destructive gales; and that others are more frequently ravaged by disastrous hailstorms. We will now endeavour to show why this should be so, and how the products of this variation are related to the general principles of the dependence of weather on the distribution of atmospheric pressure which we have already described so fully.

If we watch the actual occurrence of any local peculiarity of weather, we shall soon find out that in every instance it is the intensity, and not the general character, which is altered. For instance, two places a few miles apart may differ by 10° of temperature on a frosty morning. No local cause has formed the distribution of pressure which gives the necessary calm. That stillness has been developed by general causes, while it is local peculiarities of exposure, etc., which have enabled radiation to be so much more powerful in one place than another. Similarly, an inch of rain may fall in one place, and only a few drops in another not far distant. But if we think of the day on which this occurred, we shall remember that it was cloudy and showery naturally. The difference of actual rainfall was either due to one place catching a heavy shower which did not affect the other, or else some local peculiarity of the one, which increased the amount of precipitation that would otherwise have been induced by a cyclone of any given intensity.

These instances might be multiplied indefinitely, and it is from the observation of innumerable cases that we are enabled to lay down the general law that the primary character of all weather is given by the general distribution of surrounding pressure; the local variation modifies but never alters this general character. By this means we are able to steer our way through many intricacies of weather which would otherwise present hopeless difficulties, and to explain many phenomena which would otherwise be inexplicable. Hence we see the appropriateness of the word "variation," as applied to the modification due to local causes.

The cases which present themselves in practice are endless. Every country, every part of a country, has a set belonging to itself, and the local meteorologist has to work out the details for his own neighbourhood, just as the geologist explains the local peculiarities of his own scenery by the combination of general and local causes. Local is also like diurnal weather, in so far that the observed weather is the sum of the local variations and general causes. When the general are strong, the local are entirely masked; when the general are weak, then the local become of primary importance. We shall confine ourselves to a few examples relating to cloud, rain, and hail, so as to exemplify the general principles involved.

Local Cloud.

By far the most important and difficult source of local variation of weather is found in the development of cloud, rain, and other forms of precipitation by the influence of seas, lakes, rivers, hills, and valleys; some of these phenomena are so interesting that we propose to devote a few pages to their consideration.

We will commence with cloud, though we must remember that in most cases cloud is only undeveloped rain, and that the same cause which, when slight, gives cloud will give rain when more intense. In our diagram of cyclone-weather and prognostics (Fig. 2), we have marked cumulus-cloud in rear of the trough. In England this holds all through the year, but in the drier climate of Continental Europe it is only true during the summer months. The reason is simply that, in cold climates, there is only sufficient vapour to develop that form of cloud in summer. What we have specially to note is, that when cumulus does not form, it is not replaced by any other kind of cloud, but the sky clears without any cloud at all. For instance, no local variation could ever turn the cumulus of the rear into the cirrus-stratus of the front of a cyclone; the quantity, not the quality, could alone be changed. Similarly for rain. We shall see directly that the actual amount may vary enormously from local causes, but no peculiarity can turn the drizzling rain of a cyclone-front into the heavy, big-dropped shower of a thunderstorm, or *vice versâ*.

A very striking illustration of the influence of local peculiarities in the formation of cloud was once observed by M. Flammarion during a balloon-voyage from Paris to the Rhine. He saw one afternoon that cloud was formed over the rivers and woods, but not over the open plains. The synoptic charts for that evening show that Eastern France was then covered by a large cyclone of moderate intensity; and the explanation of the whole is, that all the air in that part of Europe was then in a rising condition, but that it was only over rivers and damp woods that enough vapour was present to condense into cloud. A more intense cyclone would have developed cloud everywhere, and rain only over the rivers and forests; another still more intense would have brought rain everywhere. In our chapter on Prognostics, we alluded to mist being formed over rivers in fine frosty weather. Here, too, we have local variation, but of a contrary nature to that which we have just considered. This example will serve to call attention to the great importance of differentiating between the various kinds of condensed vapour.

Another very common local cloud is that which rests

on or over hilltops, when blue sky covers the plains. This, of course, is due to the horizontal currents of the air being deflected upwards, and, if sufficient vapour is present, cloud is formed by condensation. The most interesting thing about these clouds is that they remain stationary as a whole, though their outlines and constituent particles are in constant motion. Their prognostic value has been already explained in a previous chapter.

Local Rain.

We shall now explain a few of the principal causes which affect the quantity of rainfall. One of the commonest and most obvious is that, when the wind which blows over water first meets the land, rain will be precipitated. For instance, in England, with a cyclone of moderate intensity and a westerly wind, rain will only fall on the western coasts and on the high ground inland. With an east wind, on the contrary, the fall will be confined to the most exposed portions of the east coast, and in a less degree to high inland stations. A similar effect is found all over the world. For instance, in Ceylon the rainy seasons on the two sides of the island are in different months, which depend on the time when each coast is exposed to the prevailing monsoon. The south-west monsoon brings rain to the exposed west side of the island, and the dry season to the east coast, which is then a lee shore. The north-east monsoon, on the contrary, first strikes the east coast, and develops abundant rainfall there; while the west coast then enjoys its dry season. A similar sequence is observed on the opposite coasts of the island of Luzon, in the Philippines, and for a similar reason.

But though the sea may assist in the development of rain under suitable circumstances, the presence of water alone will not cause rain. The rainless districts of Peru and Arabia both border on the sea-coast, but no rain falls in either country. In the former the persistent anticyclone which habitually covers that country does not form the ascensional current necessary for rain, though the air is damp enough to deposit very copious dew; in the latter, though the isobars are sometimes curved cyclonically, the rising currents never seem to be sufficiently strong or vapour-laden to produce rain.

Allied to the influence of water in supplying an abundant source of vapour, is the presence of a damp surface, such as a thick forest. The leaves of the trees retain so much moisture that the air is always damper over wood than over earth. Then, as we have just explained in talking about the development of cloud, when an ascensional impulse is propagated over the country, rain will sometimes fall over the woodlands when it would not be precipitated over the cultivated soil. For instance, in the State of Iowa, Dr. Hinrichs has shown that the amount of mean rainfall is very materially influenced by the position of the timber-line, and numerous similar cases have been recorded in other parts of the world.

But observations on this point are very discrepant. In some parts of Germany the influence of forests is said to be enormous; in India the effect appears to be less marked; while the Swedish meteorologists can find little relation between rainfall and the covering of the earth. The facts are doubtless as they have been described; and the apparent discordance is only another example of the great principle that all meteorological results are the balance of various circumstances. The difference in humidity over trees or soil will be much less in a cold country like Sweden, than under a blazing Indian sun; and a great deal will depend on whether the principal rainfall is induced by cyclones or secondaries.

Mountain Rain.

So far we have treated of water and forest as merely supplying the material for rain; now we must consider a little more in detail how hills and valleys will affect the precipitation of any current. In England, roughly speaking, if a range of hills under about fifteen hundred feet obstructs the prevailing westerly wind, the greatest amount of rain will fall on the east side of the range. This is because, though the moist current is deflected upwards on the west side, the condensed vapour is blown over the top of the hill and falls on the opposite slope. If the range is over fifteen hundred feet, then the rain cannot blow over, and the greatest rain will fall on the west side of the hills. No rule can, however, be laid down except in very general terms, for every hill and every valley has its own local peculiarities in the manner in which it develops rain with different winds. Every country, and every part of each country, must be worked out in detail on the spot.

The amount which will be deposited at different heights will also vary from a number of circumstances. For instance, on the west coast of Scotland, which is constantly exposed to south-west winds, the rainfall on the low western islands is only about forty inches in the year; while along the watershed, which forms the backbone of Scotland, the precipitation exceeds one hundred inches in many places. In Ceylon, to which we have already alluded, the fall on some coast stations does not rise above thirty-four inches, while some of the mountain stations record no less than two hundred and nine inches. The difference is readily accounted for when we consider the relative altitudes of the respective mountains and the greatly increased quantity of vapour which an air-current of 90° temperature can carry, compared with one of only 40°. These two latter numbers represent about the mean temperatures of the two countries; and while the watershed of Scotland rarely rises above two thousand five hundred feet, many of the mountains of Ceylon attain an altitude of six thousand feet.

Valley Rain.

It is this property of mountains in developing rain which gives truth to the well-known saying, "Hills draw rain." But there are two sources of rain which are intensified by valleys—the rain of thunderstorms and tidal showers. These we must now consider. In Great Britain it is a common remark that thunder-showers have a tendency to run along the course of rivers. The only class of thunderstorm which does not follow this rule is that particular kind which occurs in the winter months on the exposed western coast of Scotland and Ireland. These are certainly thunder-squalls, which belong to large cyclones, much developed by mountains. In France and other countries an immense amount of labour has been expended in tracking thunderstorms, as we have already mentioned in a preceding chapter. Though the storms as a whole travel in a north-easterly direction for short distances, the course of rivers is found to exercise a very powerful influence both on their path and still more on their intensity. Forests and hills also modify the development of thunderstorms to a less extent, so that we may conveniently consider them all together.

Localization of Hailstorms.

But first we may give an example of the actual facts. Since hail may be considered as the most intense form of a thunderstorm, we have given in Fig. 59 a reduction of a chart illustrating the distribution of hail in the French Department of Loiret.

The river Loire will be readily recognized running across the diagram from right to left, as well as some of its smaller tributaries. A well-known conventional symbol marks the limits of the forest of Orléans, while the small, round points indicate the number of years in which any commune has been attacked by hail during the thirty years 1836–1865. The scale of miles shows what a small area we have to deal with; but see what a difference in the number of hailstorms. In the town of Orléans, on the river Loire, sixteen destructive hailstorms have been recorded; and at the village of Jangeau, a little higher up the river, twelve serious falls. Yet within a couple of miles of both places, other communes, such as Clery, have not been visited by more than two or three storms. Then observe the influence of the mass of forest to the north of

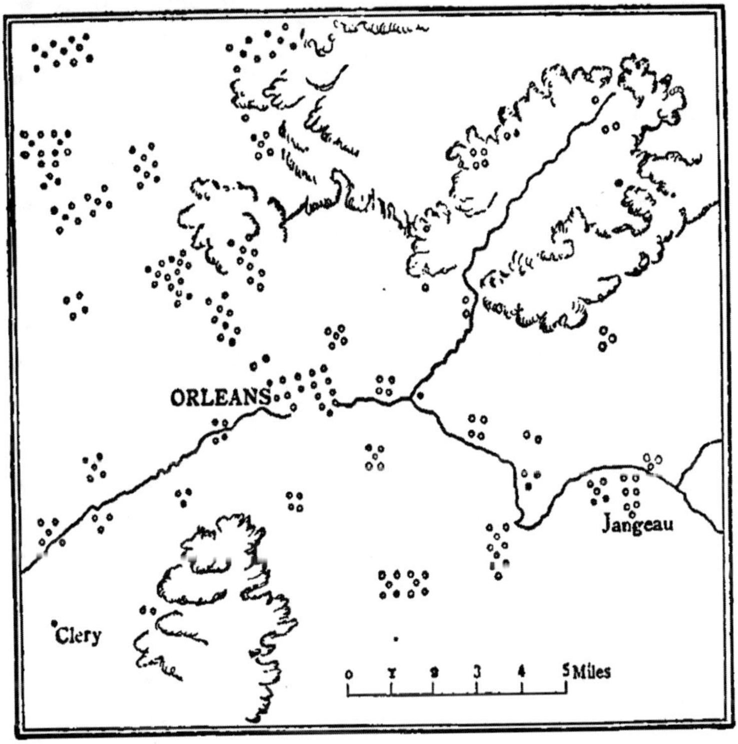

FIG. 59.—Localization of hail in Loiret. The points indicate the number of years in which any commune has been attacked by hail during the thirty years 1836–1865.

the river. All the villages to the west of the forest area have a large number of points, while those inside the forest enjoy almost complete immunity from destructive hailstorms.

The forest seems to act, in fact, as a breakwater, against which the violence of the storms expends itself for a time, till it can gather fresh force. Almost all the French observers are agreed as to the origin of this development and protection. We must recall the fact which we mentioned in our chapter on thunderstorms, that the wind-circulation of thunderstorms is always confined to the low strata of the atmosphere only.

Hail is produced when two clouds are superimposed at a certain distance. A storm is never isolated. The ordinary French ones are often formed by partial derivations taken from the south-west winds, and occasioned by the passage of cyclones. When the thick mass of cloud which marks the existence of a storm meets a valley, the lower clouds are diverted from the general route; a portion follows more or less exactly the contours of the valley. Thus it happens that the clouds passing at a great height from the ground cross those which, entangled in the valley, have been deflected by the successive bends of the river into a direction different from the south-west. It is then that hail falls.

In the same way, when the lower clouds meet such an obstacle as a forest, or even a mountain, eddies are formed. The masses of cloud come back on themselves, and seem to be repelled and dispersed by the forest. When the clouds have succeeded in passing the obstacle, their force is exhausted, and they only precipitate rare or inoffensive hail, and do not regain their intensity for some time afterwards.

From this we can readily understand the manner in which hills and valleys develop their respective rains.

In a large, deep cyclone, hills deflect the moist currents upwards on a large scale, and the greatest rain falls in the mountains. In shallow secondaries, with slight general wind-force, the rivers and forests give rise to local eddies, which for some reason develop the precipitation of rain, and especially of hail.

Tidal Showers.

Tidal showers are of very little practical importance, but they may advantageously be mentioned as belonging to the class of rain which hugs the valleys, and not the hills. These showers are so called because they are brought up by the tide, either along the coast or up tidal rivers. How the rising water should develop rain, we cannot explain, but the character of the influence is very obvious. On a cloudy day, when showers or heavy masses of vapour are flying about, it is frequently observed that after the tide turns to rise, and the stream is running upwards, the weather begins to get worse, so that what was merely a mass of cloud before will now precipitate rain. This rain is quite local, and does not extend far from the river-banks.

In calm weather, a wind also often comes up with the tide, or if the flow of the tide assists the general direction of the wind, the latter will be much increased in force and gustiness. If the day is really fine, of course the tide will not bring up any rain, though it may modify the wind.

From this description, the general nature of tidal action on weather will be sufficiently obvious. For some reason or other the rise of the tide increases the intensity of the existing system of weather. If this is tending towards precipitation, the tide will give just the last impulse which is required, and rain will fall. If the ascensional impulse is strong enough of itself, it will rain independently of the tide; and if the natural impulse is downwards, as in an anticyclone, no tide is sufficient to invert the general character of the weather and cause rain. Tidal influence on weather is found all over the world. Professor Hazen has found a marked increase of thunderstorms with a rising as opposed to a falling tide in the United States; and the author has observed a well-defined tidal variation of the trade wind in tropical Fiji.

We can, therefore, sum up the contents of this chapter very easily. All over the world local influences modify, but do not make, the general character of the weather. When the latter is weak, local weather may be the prominent feature of the climate of any place; when it is strong, then local influences may be entirely obliterated.

CHAPTER XI.

DIURNAL VARIATION OF WEATHER.

In this chapter we propose to explain how to collate the variations of weather that are found in many places to depend on the hour of the day, with the great principles of the relation of weather to the distribution of atmospheric pressure which lie at the bottom of all modern meteorology. In many places the direction of the wind changes regularly at certain hours, or cloud and rain gather at the same time day after day; how can all this be reconciled with the laws of the dependence of wind on gradient, and of weather on the shape of isobars?

The cases which arise in practice are endless. Every country, every season, has its characteristic diurnal weather, and a complete account of these variations in some climates would more than fill the whole of this volume. We must, therefore, content ourselves with a statement of the general principles which are found by observation to regulate all diurnal variations, and with illustrations of a few typical examples from various parts of the globe.

Independence of Diurnal Variations and General Changes.

The great principle which underlies all diurnal weather is that diurnal variation modifies but never alters the general character of the weather, which is determined by the distribution of surrounding pressure. In England the amount of cloud or rain in a cyclone will vary at different hours, but the kind of cloud and quality of the rain will never be altered. In like manner, the land and sea breezes at Bombay will veer or back with the sun, but the general character of the wind due to the monsoon of the season will never be lost. This law of weather not only enables us to explain many phenomena of weather which would otherwise present a chaos of discordant observations; but it also serves to guide our research into the great problems of weather-forecasting. When once we know that we may safely neglect all considerations of diurnal variations when we wish to study the motion of depressions and the consequent changes of weather, our task is thereby much simplified. If we were to discuss the statistical values of meteorological elements, we should find that all the voluminous results which have been obtained by the method of averages are of no use in forecasting, and that the diurnal values of wind or rain which have thus been obtained have nothing to do with weather-changes.

Diurnal Temperature.

In this independence of the general changes of weather, diurnal are very like local variations; but there is one important difference—that diurnal variations introduce us for the first time in this book to the consideration of the true nature of all meteorological periodicities. From the variations which run through their entire course in one day, we can readily pass to those whose period is one year, or even a longer cycle.

We shall therefore commence with an account of the nature of the diurnal variations of heat, as that is the most obvious of all meteorological phenomena. It is evident to all the world that, whatever the temperature of the day or season may be, the nights are in a general way colder than the days; this is the ordinary instinctive idea of diurnal variation. We also know almost by instinct whether it is a generally cold or hot day; and equally instinctively we allow for the difference of day and night. Put into the formal language of ordinary meteorology, the general heat or cold of the day is expressed by the number which gives the mean temperature of the day; while the diurnal variation or range is given by the numbers which denote how much the thermometer was above or below the average at each hour. In a climate like that of England, most people have a vague idea that sometimes in winter the weather gets warmer in the evening, after a white frost in the morning; but when we come to examine self-recording thermograms, we find that irregular changes of temperature are much more common than is usually supposed. We have already alluded to this subject in our chapter on Meteograms, but we give here, in Fig. 60, the thermogram at Kew, December 7 to 9, 1874—that is, for the same period as the meteogram and charts in Figs. 25, 26, and 27; and in addition to this we give in Fig. 61 the mean annual curve of diurnal variation of temperature at Kew, as deduced by Mr. Eaton. The point which most concerns

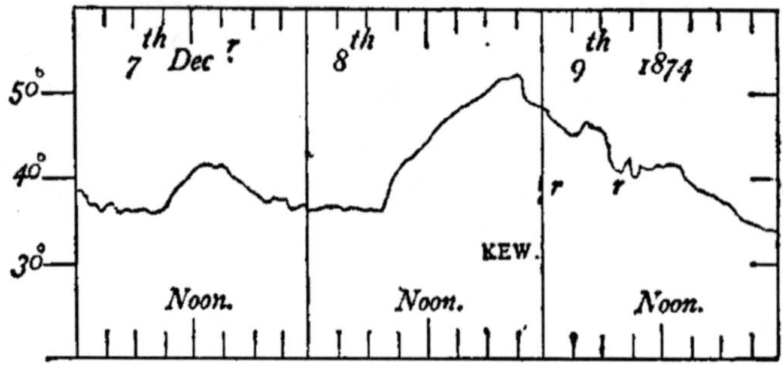

FIG. 60.—Thermograms.

us here is to understand exactly what the significance of the curve of mean temperature is. The thermographic trace for three days at Kew, in Fig. 60, shows very fairly what temperature-changes are really like from day to day. In spite of endless irregularities, there is a general tendency in the first two days for the hottest time of the day to occur in the afternoon, and the coolest in the early morning; while on the third day there is little trace of diurnal variation, but a steady general fall of temperature due to general causes.

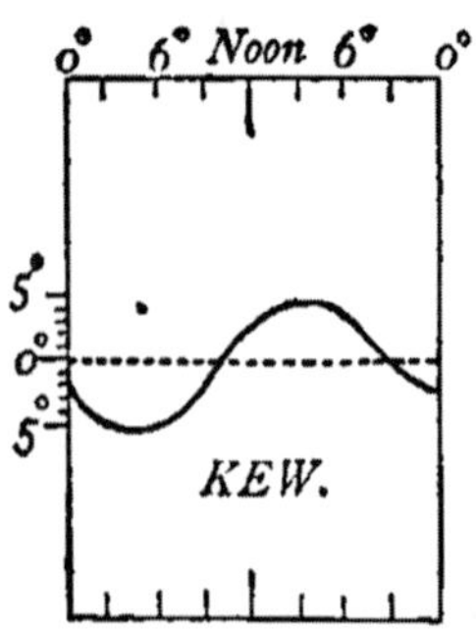

FIG. 61.—Mean diurnal range of temperature.

When the mean temperature for every hour is taken on a great many days, the irregularities balance out, and the daily tendency only is reflected in the mean curve of diurnal range.

If we lived in a vapourless climate, the sun would

impose every day a similar record on the thermograph, only varying a little in amount according to the season. But in practice the sun has a daily struggle with the wind and cloud. Some days a shift of wind to the south in the afternoon will cause the thermometer to rise steadily as the sun goes down, and midnight may be hotter than noon; on other days a dense layer of mist will completely shut off the influence of the sun's rays, and the instrument will leave a straight horizontal line as its record of temperature-variation for twenty-four hours.

All that the mean curve of temperature signifies is, that in a general way, allowing for all sorts of irregularities, there is a diurnal solar influence, which has its greatest and least values at such and such hours. But we must most carefully avoid two conclusions: first, that the mean temperature represents any abstract entity, called mean diurnal range, which might be applied as a correction to the observed temperature at any hour so as to deduce the mean temperature of the day; and, second, that because we do not see a diurnal variation on the trace for every day, therefore there is no such thing as solar diurnal influence. Because the thermogram for December 9 (Fig. 60) shows no diurnal *maximum* and *minimum*, that does not prove that there is no such thing as diurnal variation at all. The importance of this last conclusion will be evident in our next chapter on Cyclical Variations of Weather.

What more immediately concerns us now is to note how diurnal affects general heat. This can be better accomplished by trying to recollect the history of any particular day's weather than by the inspection of diagrams. If we think of any day, we shall remember that if the wind did not shift, the general character of the heat did not alter. Suppose, for instance, that we had been in a large cyclone with a north-west wind, which lasted for the whole twenty-four hours. We should have known instinctively that it was a cold day for the season, and though there would have been a very considerable difference between the temperatures recorded by night and by day, the quality of heat which belongs to north-west winds would have remained the same. But suppose that, after a sharp white frost in the morning, about midday the sky had begun to thicken and the wind to back to the south, then, as before mentioned, the temperature would rise while the sun was going down, but every one would have recognized that the quality of the heat and the general character of the weather had changed. We thus see the correctness of the phraseology which calls such changes general, which have their origin in the great movements of atmospheric pressure with wind-shifts, and those changes variations which are imposed on the general character by the sun's daily influence. In fact, we realize the great principle that the diurnal variations are superimposed on the general changes, but never alter the character of the latter. Temperature, like every other element of weather, is the balance of the general changes and diurnal variations; so that when the general are strong the diurnal are masked, when the former are weak the latter are predominant.

Diurnal Cloud.

From the comparatively simple nature of diurnal heat, we must now turn to the far more complicated variations of fog, cloud, and rain. The simplest case of the diurnal precipitation of vapour is found in the regular formation of valley or river mist in fine weather. In settled climates, at certain seasons, the sky is always blue by day, but after dark, fog or mist begin to form in the low-lying ground, from the influence of nocturnal radiation. By sunrise the valleys will be filled with mist, which rises to such a uniform level that, viewed from a height, the hollows look as if they were filled with water. After the sun is up, the vapour gradually rises and disperses, till the sky resumes its usual blue appearance. The general anti-cyclonic character of the weather is the same throughout, but the day impresses a variation on the face of the sky.

Diurnal weather-variations are so intricate, and vary so much, not only in different countries, but at different seasons, that we can only give a few illustrations of general principles. Every one of the seven fundamental shapes of isobars has a type of diurnal weather peculiar to itself. Only that of the two great shapes has been worked out for Great Britain by the author.* He has shown that in their diurnal variations cyclones and anti-cyclones present the same antithesis as they do in all their other special characteristics. The broadest features of the diurnal variation in a cyclone is that, starting from the early morning, the amount of cloud and the

* *Quarterly Journal of the Meteorological Society*, London, vol. iv. p. 4.

general severity of the weather gradually increase till about 2 p.m., and then gradually decrease till past midnight. In anticyclones, on the contrary, a misty or cloudy morning is followed by a great diminution of cloud as the day goes on, while later on, in the evening, mist and cloud are sometimes formed again.

If, besides these broad features of diurnal variation, we consider some of its more minute changes, we observe that, in cyclones, a cloudy morning often has a short break about 10 a.m.; that the weather then becomes much worse, but has a marked tendency to clear up again about 4 p.m. In anticyclones, on the contrary, a clear morning at 4 a.m. is frequently very cloudy at 10 a.m., after which the cloud again decreases till about 4 p.m., when more cloud or mist are often again formed, and last till quite late in the evening. We find, in fact, not only traces of a semi-diurnal variation, or of one which runs through its course twice in a day, but also that the intervals of this variation are obviously connected with the hours of diurnal *maxima* and *minima* of pressure.

The increase of cloud during the day in cyclones may be generally described as an accession of intensity which accompanies the diurnal increase in the velocity of the incurving wind. Then, as more air is being poured into the centre of the cyclone, the ascensional currents must also be stronger, and therefore more cloud will be formed. In anticyclones, the morning mist has already been shown to be due to radiation, and the marked clearing of the sky during the day must be due to the increased strength of the descensional currents near the centre, caused by the increased velocity of the outcurved wind during the day. From this we can easily understand how two different shapes of isobars can have such very different diurnal variations. In one, the influence of the sun modifies a rising current; in the other, a descending one.

But whatever modification the diurnal variation may impose on a cyclone or anticyclone, the general character of either is never lost. No diurnal variation can make the cirrus clouds in front of a cyclone either like the cumulus in rear, or like the clouds formed from the rising mist of an anticyclone. The variation is, as it were, a modifying influence superimposed on the more general features.

Diurnal Rain.

The diurnal variation of rain is one of the most difficult questions in meteorology. Not only does the variation differ in each shape of isobars, but there is a tendency of small secondaries to form at particular hours; and moreover, during the winter months, there is a marked tendency of large cyclones to come in from the Atlantic with increased intensity during the night. Besides a diurnal variation of the intensity of a rainy shape of isobars, there is a diurnal period both of the formation and motion of cyclones.

The error which we have most carefully to avoid in treating this branch of the subject, is the supposition that all rain is cyclonic, or that the diurnal period of thunderstorms and secondaries is the same in every country. From this it is evident that if one shape of pressure-distribution predominates in summer, and another in winter, then the two seasons will have a totally different type of diurnal rain-variation.

We can get a most striking illustration of these principles from the tropics. At Calcutta Mr. H. F. Blanford finds* that the weather is divided into three seasons:—

The rains, June to October, when the diurnal frequency curve of rain begins to rise soon after midnight to a small *maximum* about 6 a.m. and a small *minimum* about 8 a.m. Then the frequency rises rapidly to its principal *maximum* at 2 p.m., and falls quickly to the principal *minimum* at 1 a.m. The mode of the formation of the rain-cloud of the summer monsoon is essentially cumulus.

The hot season, March to May, when the diurnal epoch of *minimum* is not very distinctly indicated, but would appear to occur about sunrise. There is, however, little variation from midnight up to 9 or 10 a.m.; and after this, only a slow rise up to 2 p.m., when the increase becomes more rapid. About two hours before sunset, there is a sudden rise of about fifty per cent., and the hour of *maximum* raininess occurs between 7 and 8 p.m. Compared, however, with the *maximum* of the rainy season, it is very small. This very striking feature of the hot season is due to evening storms, known as north-westers, which are so called because they commonly originate in the north-west, and are probably connected with the diurnal variation of wind near the coast.

Lastly, the cold season from November to February. In this, falls of rain are pretty evenly distributed throughout the day, with a decided diminution during the two or

* Asiatic Society, Bengal, xlviii., part ii., 1879.

three hours before and after midnight. These seasons are, of course, associated with different types of pressure-distribution. The rainy, cold, and hot seasons belong to the periods of the south-west and north-east monsoons, with an intermediate period respectively. When these are all combined into a yearly curve, the result is a curve which gives the true variation of no season. In this case, as the rainy season curve is very pronounced, and that of the two other seasons much less marked, the annual curve differs but little from that of the rainy season, though some of the minor flexures are altered. Under other conditions the mean annual curve might be very different from that of any one season.

In some countries, where land and sea breezes are tolerably constant, rain often falls at the turn of the wind, but the details vary indefinitely; and in many parts of the tropics, inland as well as on the sea-coast, thunder-storms form regularly at the same hour every day. Inland this does not usually occur before 2 p.m., and usually later; but no rule can be laid down. These storms are, of course, totally non-isobaric.

All over England the mean diurnal curves of rain are so irregular that they do not show any real variation; for there are so many kinds of rain in that country—each with its own variation—that the curve of all mixed up together has no physical significance at all. In Prague, Professor Augustin finds three types of diurnal frequency for winter, summer, spring and autumn, respectively; while in many parts of Europe, and in Japan, there is a tendency to develop morning and evening *maxima* of rain, with day and night *minima*. Every country has its own peculiarities; and each set must be explained on its own merits.

In our detail of the cyclonic variation of cloud, in England, we pointed out that the diurnal variation does not alter the general character of the sky or cloud, or we might say of the weather as a whole. The same great principle holds for every other shape of isobars and for every other climate. The diurnal variation of rain in Calcutta during the rainy season is enormous, but at all hours the general character of the south-west monsoon is always the same. Similarly, during the cold weather of the north-east monsoon the diurnal variation is only a modification, not a real change of weather.

Diurnal Wind.

In our chapter on Meteograms we have partially explained the general idea of diurnal wind-variation, but we now wish to give some additional developments of the subject.

Diurnal Velocity.

The commonest feature of diurnal wind in most places, temperate as well as tropical, is an increase of the velocity from daybreak to about 2 p.m., and then a decrease to its lowest point about 4 a.m. Besides this, there is in many places a smaller series of variations whose turning points occur about the same time as the *maxima* and *minima* of diurnal pressure. For instance, in Great Britain, the principal *maximum* is about 2 p.m., and the principal *minimum* about 4 a.m. But in addition to this, there is a small *minimum* about 10 p.m., with a small *maximum* between that hour and the great *minimum* at 4 a.m., besides a well-marked *minimum* about 10 a.m., just at the hour when cyclones and anticyclones develop their most characteristic difference of diurnal formation of cloud. A similar tendency is found all over the world, both north and south of the equator. The details vary indifferently, and cannot be said to have been fully worked out in any one country.

We can, however, safely say that the variations are all of the diurnal type, and have nothing to do with the character or amount of wind-velocity, which depends on the distribution of surrounding pressure. Just as with weather, the general character of wind and the relation to gradient depend primarily on the isobars; the diurnal variation, in spite of its great complexity, is purely secondary.

Buchan has made the very important observation that the diurnal variation is almost nothing over the sea, when away from the influence of land, and he has also connected this with the fact that the diurnal variation of temperature is very small over the sea compared with that over land, so that in some way the diurnal amount of wind-velocity appears to depend on the temperature of the floor over which it blows.

Many of the details of wind-variation are both interesting and puzzling. In some places the wind falls about noon, probably from some local influence; and in Great Britain, Ley has shown that the diurnal variation of velocity is greater with west than with east winds. This again coincides with an observation of Hamberg's, that the stronger the wind, the greater the amount of diurnal increase of velocity, because as a rule in England west winds have a higher velocity than east ones.

So far we have only thought of the wind on the earth's surface, but on high mountain peaks the variation of wind velocity is almost exactly the opposite; for there the *maximum* is in the early morning, and the *minimum* about or just after noon. The general speed of the wind is, however, much greater at high altitudes than at low levels. This principle appears to hold equally in both hemispheres.

Diurnal Direction.

The diurnal variation of direction is less marked than that of velocity and much more difficult to detect. In the northern hemisphere, however, there is a well-defined tendency to veer a little in the morning, and to back again in the afternoon; so that the times of greatest veering and backing correspond to the hours of greatest and least velocity. That is to say, if the general direction of the wind is from the west, it will be a little more from the north-west by day, and a little more from the south-west by night; the greatest northing being about 2 p.m., and the greatest southing about 4 a.m. There are also traces of a semi-diurnal variation exactly similar to that which we described under the head of velocity; but we cannot give complete details for any one place.

On mountain-tops the daily oscillation of the wind is on a contrary system, for there the wind backs by day and veers towards evening. For instance, a generally westerly wind will back towards west by south till the afternoon, and then veer towards west by north at nightfall. We demonstrated in our chapter on Clouds, under the heading of "Vertical Succession of Air-Currents," that in the northern hemisphere successive vertical strata of wind came more and more from the left of an observer standing with his back to the wind on the surface. If, then, the surface current veers, and the upper ones back during the day, the result will be that, both in cyclones and anticyclones, the difference in direction between the upper and lower currents will be greater by night than by day.

In the southern hemisphere we have not a sufficient number of observations to enable us to generalize on the nature of diurnal wind-variation; but as far as they go they point to an exactly similar law to that which holds in the northern hemisphere. The surface winds veer, and the upper currents back with the course of the sun. But observe that the course of the sun is opposite in the two hemispheres, so that a westerly surface wind would veer towards north in the northern hemisphere, and towards south in the southern hemisphere. As some may prefer to see the laws of diurnal wind exhibited in their relation to absolute direction, as given by the hands of a watch, we may state these results thus—

		Forenoon.	Afternoon.
Northern hemisphere—	Surface	With watch-hands	Against.
" "	Hilltops	Against "	With.
Southern hemisphere—	Surface	Against "	With.
" "	Hilltops	With "	Against.

The following is a fair illustration of the nature and amount of diurnal wind-variation, both of velocity and direction, as ordinarily observed in Great Britain. In Fig. 62 we give a copy of the anemographic record at Kew, near London, for the three days, August 6th to 8th, 1874.

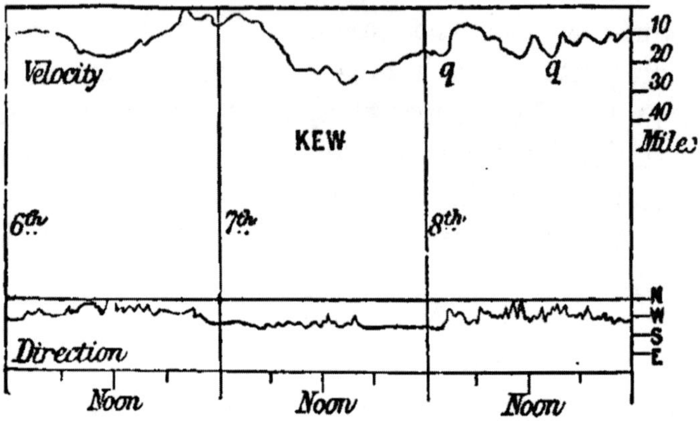

FIG. 62.—Anemographic curves for Kew, August 6th to 8th, 1874.

The synoptic conditions for these three days were the commonest in that country. A series of large cyclones of moderate intensity were passing to the north of Great Britain, so that, although there was a good deal of cloud and wind, there was not the marked shift of wind which would occur if the cyclone centres had passed nearer the station. Taking the velocity first, in the first two days the tendency of the wind to rise during the day is very obvious; but on the third day the ordinary variation is completely masked by violent squalls. Two of these occurred at the times marked *q* on the diagram. The smaller semi-diurnal variations are also almost completely obliterated; they are only observed in calm, summer, anticyclonic weather.

Then, looking at the direction-traces, the general

westerly direction of the wind due to the isobars is sufficiently obvious, but superimposed on that we find every day an irregular tendency for the wind to veer a little towards the north-west during the day, and to back again during the night. We also see another feature of British winds, viz. the increased gustiness by day compared with the night. This is shown by the more irregular trace during the day hours.

From a simple case of this sort we can readily see both the amount of diurnal variation as well as the manner in which it can easily be masked. When the general features of the weather are feebly marked, then the diurnal variations are strong, and may be the prominent character of the day. This is common in many tropical countries, especially in those which are habitually covered by anticyclones. In variable climates, like that of Great Britain, on the contrary, the diurnal variations are only obvious in the finest settled summer weather; and in winter, when the general changes are very intense, the diurnal features are often completely lost.

We may then lay down as a general rule that the prominence of the diurnal variations of weather are a measure of the settled or unsettled character of the climate of any place.

Any attempt to discuss all the details of diurnal wind, or the different theories which have been suggested, with more or less probability, to account for it, is beyond the scope of this work. All that we wish to do is to give a clear sketch of the general nature of diurnal variation, and especially of the manner in which it is subordinated to the great laws of dependence of weather on isobars.

General View of the Subject.

It is most interesting to note the unity which runs through the whole class of diurnal variations. The principal *maxima* and *minima* of temperature, wind, and partially weather at 2 p.m. and 4 a.m. respectively, are strictly analogous to each other, while the semi-diurnal features of wind and weather are analogous to the diurnal variation of pressure. The latter, to which we have scarcely alluded, has two *minima* at 4 a.m. and 4 p.m., and two *maxima* at 10 a.m. and 10 p.m. respectively. The single diurnal variations are undoubtedly due to the direct influence of the sun's heat; but the question how an influence such as that which runs its course only once in the twenty-four hours can induce a variation which has a semi-diurnal period, has, up to the present time, baffled the skill of meteorologists. It is, however, perfectly certain that no one is the cause of the others; all are equally the products of the same influence, and no comprehensive theory of diurnal variations will ever be complete which does not explain and co-relate all together. When we look at a series of synoptic charts for several consecutive days, we see that many cyclones go on their course often for two or three weeks, quite independent of diurnal changes. We may, therefore, perhaps suggest the following broad view of the relation of diurnal variation to the general character of weather. The whole atmosphere is circulating between the equator and the poles. Sometimes this flow of air takes the form of eddy known as a cyclone, sometimes that known as an anticyclone, and almost always one of the seven fundamental forms of circulation. Every day, as the sun rises and sets on this system, he impresses either directly or indirectly a series of complex variations on every meteorological element, but does not change the intrinsic nature of any form of circulation.

The results of this chapter may therefore be summed up as follows. In every part of the world the diurnal variation is superimposed on the general character of the weather, which is due to the distribution of surrounding pressure. The resulting weather is the balance of the general character and diurnal variation; the prominence of the diurnal is a measure of the settled nature of the climate of any place.

All over the world there is a tendency to form both a single diurnal variation, which varies only once in the twenty-four hours, and a semi-diurnal variation, which has two *maxima* and two *minima* in the same time. The origin of the first is undoubtedly the direct action of the sun; that of the latter cannot be at present explained. No diurnal variation has any effect on general weather, and can be neglected in all questions which relate to forecasting general changes. This independence is one of the most important principles in meteorology.

CHAPTER XII.

ANNUAL AND SECULAR VARIATIONS.

SEASONAL APPEARANCE OF THE SKY.

THE term "seasonal variation" is used in a twofold sense. In the simpler case, it refers to the minute differences in the appearance of the sky which are found at various seasons in cyclones, etc. For instance, the rear of a cyclone does not form cumulus cloud in the dry winter months of Continental Europe; only blue sky is seen. In damp England, cumulus is formed at all seasons; but is much denser and more strongly marked in summer than in winter. In like manner, a secondary which would develop thunder in summer in Great Britain would only produce heavy rain in winter. In this way seasonal is exactly analogous to diurnal variation, for it modifies but never changes the general character of the weather. The intensity alone is ever altered.

RECURRENT TYPES OF WEATHER.

But of far more importance is that form of seasonal variation which applies to the occurrence or recurrence of similar weather about the same date every year. The nature of recurrent weather in the temperate region of variable pressure may be best illustrated by looking at the connection between the variable European types and the regular annual changes which take place in the tropics. In most equatorial and tropical climates there are only two or three seasons, which correspond to two or three positions of the equatorial low pressure and tropical belt of anticyclones. The monsoons of the Indian Ocean are the most striking and best known instances of weather that recurs at the same season of every year.

The English word "monsoon" is, in fact, derived from an Arabic word meaning "season." But in the regions which lie between the tropical and temperate zones the author has found that there are recurrent periods, intermediate both in their duration and the certainty of their return to the monsoons of India and the recurrent spells of European weather. One of the best known of these is the "Khamsin" (the fifty days), a hot, sandy south-east wind which blows regularly in Egypt from the end of March and through April for about fifty days. Klunzinger has given a whole series of persistent and recurrent types for the whole year in Kosseir, on the Red Sea, about one hundred miles south of Suez on the Egyptian side. The following may be considered a list of the chief recurrent periods of weather in Great Britain and North-Western Europe generally.

February 7 to 10.—A spell of cold weather, associated with the northerly type. This is the first of a series of six cold and three hot periods discovered by A. Buchan. He also noticed that during the cold periods the pressure was higher to the north of Scotland, and lower to the south, and that during the warm periods pressure was higher over Scotland than in places to the north. This means that the cold periods were the result of the occurrence and persistence of either the northerly or easterly types of weather. We have been unable to find any allusion to this spell in European weather lore.

March.—The proverbial east winds of this month are mostly due to the northerly type of weather. The occurrence of equinoctial gales about the 21st of the month is almost universally believed. It is, however, a curious fact, as has been pointed out by R. H. Scott, that the records of the British observatories give no decided indications of exceptionally strong winds at either equinox. Whether equinoctial gales really occur in the Mediterranean, and the idea has been carried from thence to England by the monks, or whether the weather in Great Britain might not be more properly called broken than stormy, we cannot say. The author, however, rather inclines to the latter view; for it is almost impossible to believe that an idea which has obtained such universal credence can be altogether destitute of some real foundation. The difficulties of settling a question like this bring forcibly before us the uncertainty of any numerical estimate of climate or weather.

April 11 to 14.—A cold spell; Buchan's second period, which he has identified with the popular "weather saw" of the "borrowing days."

May 9 to 14.—A cold spell; Buchan's third period. This is the most celebrated of the cold periods, as it occurs over the greater part of Europe. A good many sayings connected with it are found in many European prognostics, such as those relating to the frost saints. This period is of some interest on account of the strange theories which have been propounded to explain the origin of the cold. One of the most popular has been the idea that about the middle of May the earth encountered a stream of meteors which were so numerous as to act like a cloud of dust and cut off some portion of the sun's heat. We need hardly say that such an occurrence would diminish the temperature all over the world, and that there is nothing to give countenance to this. Besides, the passage of the sun's rays through such a stone-strewed space could not fail to give rise to some kind of blur of light round his disc, as when he shines through big drops of condensed vapour. Nothing can be more certain than that this cold period is usually due to the setting in of a spell of the easterly or northerly type over Europe. At any other time of the year the same types bring similar weather.

June.—A cold spell in the second or third week is associated with the northerly type.

June 29 to July 4.—A cold spell; Buchan's fourth period. Curiously enough, we have been unable to find any reference to these thermometric periods in the weather lore either of Great Britain or of any other part of Europe.

July 12 to 15.—A warm period; Buchan's first.

July 15.—St. Swithin. The popular legend of this saint, and other rainy saints like St. Médard, receives an easy explanation from synoptic charts.

August 2 to 8.—A wet period; the "Lammas floods" of Scotland.

August 6 to 11.—A cold period; Buchan's fifth.

August 12 to 15.—A hot period; Buchan's second. No prognostics are associated with these two latter periods.

September.—The easterly and northerly types are rare during this month; the gales or broken weather at the equinox are almost invariably of the westerly type. About the 30th a fine period is experienced for a few days—the "Indian Summer" of North America.

October.—About the second or third week a spell of the easterly type of moderate intensity is common.

October 18.—A fine quiet period about this time—"St. Luke's Summer." This and the other summers which occur at this season have sometimes been stated to be due to the liberation of heat during the condensation of vapour, and formation of ice, which begins to take place on a large scale in the polar regions soon after the autumnal equinox. According to this theory, the opposite phenomenon of cold in April and May is supposed to be caused by the absorption of heat due to the melting of ice. Both ideas are purely fanciful. The spring cold we have already explained; the autumn summers are due to the recurrence of tranquil periods at that season.

November 6 to 12.—A cold spell; Buchan's sixth, associated with the northerly type. The 11th is "St. Martin's Little Summer," popularly considered in the Mediterranean to be a period of warm, quiet weather.

December 3 to 9.—A warm period; Buchan's third.

The general explanation of all these periodicities is identical. They all depend for their origin on a tendency of certain types of pressure-distribution to recur about the dates just mentioned. The cold periods all require the presence of the northerly or easterly types; the warm periods, either of the southerly type in winter, or of anticyclones in summer; while wet or broken periods indicate the recurrence of intense cyclones of any type.

Returning to our old illustration of a globe surrounded by a circulating atmosphere, we can readily suppose that at the same date every year, when the sun is in the same place, the motion of the air will tend to reproduce the same kind of eddies in the same localities.

Value in Forecasting.

But now we have to consider how the knowledge of these recurrent periods can be utilized in forecasting. In our last chapter on Diurnal Variations we called attention to the nature of the daily period of heat. In this, the most obvious of all meteorological phenomena, we found that though there is a powerful heating influence present every day, still that other causes are sometimes so powerful as to obliterate or invert the action of the sun. As a consequence of this we cannot affirm absolutely that the night will be colder than the day, though, of course, such is generally the case.

If we were to attempt to forcast the heat at any hour by reference to the mean curve of diurnal range, we should sometimes give most erroneous forecasts; if, on the contrary, we looked at the chart for 8 a.m., we could often safely predict that the ensuing night would be warmer than the day. From the temperature diagrams which were there given, we also drew the important inference that, because we do not see a diurnal range every day, we must not infer that there is no such thing as a diurnal solar influence. If, then, such a powerful influence as the direct rays of the sun can be so easily masked, we can readily understand that a weaker influence, such as the declination of the sun on any particular day, can readily be obliterated. We can safely say that the change in the altitude of the sun is of secondary importance, because we see every day great changes in the distribution of pressure, which are certainly in no way related to the seasonal change in the declination of the sun. We need not, then, be surprised that the types of heat or cold do not recur absolutely every year, only that there is an undoubted tendency to do so. When once we have realized this, we can easily understand the following statement of the use of a knowledge of recurrent annual types in forecasting.

A forecaster is not justified in saying that any period will occur absolutely; still, when about the time of its usual recurrence the synoptic charts show signs of the expected type, then the forecasts for a few days ahead can be issued with greater confidence. For instance, suppose that about the 6th of November the charts begin to show traces of the northerly type, then—but not before—there would be good grounds for saying that a period of cold weather, which usually occurs at this season, has already set in, and may be expected to last for five or six days. The forecaster is thus enabled to issue a much longer forecast than he can as a rule safely attempt.

Cyclical Periods.

By these we mean periods which run through their whole course in any time other than a day or a year. Many investigators have thought that they have detected traces of periodicities of temperature of about twelve days and of 25·74 days, the latter apparently connected with the time of the sun's rotation. Others have endeavoured to detect periodicities of rain or heat for longer epochs, especially one of 11·1 years, which would coincide with a period of sun-spots. As this is the one of most importance and greater interest, and as a discussion of it will serve to illustrate the whole nature of periodicities, we shall confine our attention to a short notice of this cyclic period only.

Sun-spots and Weather.

Ever since the year 1775, we have more or less complete records of the relative extent of black spots on the sun's surface. These records show a most unequivocal recurrence of sun-spot *maxima* at intervals of about eleven years; but the actual amount of surface covered at each *maximum* is very irregular. In the lower part of Fig. 63 we give a reduction of diagram which shows the relative extent of black spots on the sun as plotted by Professor Balfour Stewart. If we were to draw over this curve another which showed the mean daily range of magnetic declination for the same year, we should find that there was an unmistakable similarity between the two curves, and that both the times and magnitudes of the *maxima* and *minima* agreed wonderfully well.

In like manner a curve which showed the number of auroras observed in each year would also show a striking likeness to the curve of sun-spotted area. This curve is not so valuable as that of magnetic declination, because auroras cannot be seen in cloudy weather, while magnets

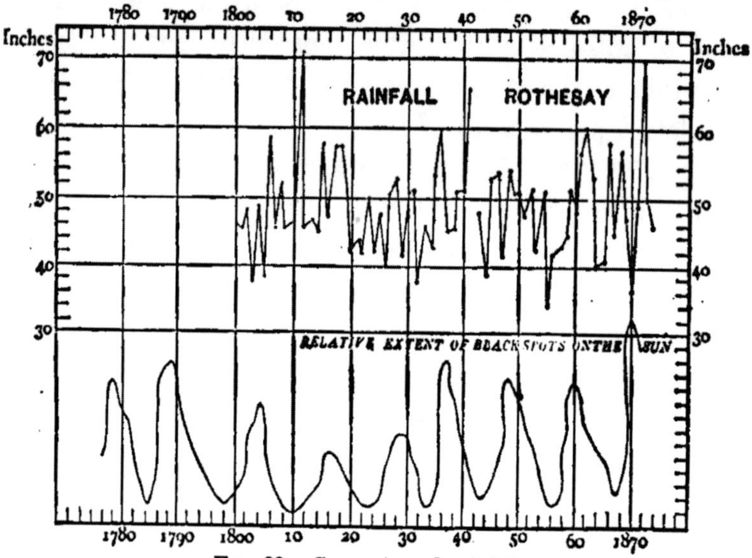

FIG. 63.—Sunspots and rainfall.

can always be observed. As these curves undoubtedly connect the state of the sun with one physical terrestrial phenomenon, and also with another half-physical, half-meteorological appearance, there would be no inherent improbability in the existence of a relation between sun-spots and weather.

Such a relation would be on quite a different footing to any quasi-astrological idea of a connection between the sun, moon, or stars, and weather-changes.

Many investigators think that they can trace some kind of connection between the amount of rainfall and sun-spots; others see a connection between the years of *maximum* sun-spots and the frequency of cyclones in the Indian Ocean; while some find that marine casualties and commercial panics or crises appear to follow a cycle closely corresponding to that of sun-spots.

One great difficulty in deciding whether there is a real periodicity in rain or storm statistics arises from the very irregular curves with which we have to deal. All the curves, on which it is sought to base the supposed connection between sun-spots and weather, have been so far smoothed, that it is difficult to say what the resulting curve really signifies and how far true deductions can be made from it. This will be better realized by an example. On the upper half of Fig. 63 we have, therefore, plotted the annual amount of rainfall at Rothesay, in Scotland, for eighty years, over the curve of sun-spot extent in the lower portion of the diagram. Both curves are purely the result of observation and have not been smoothed in any way. The reader can, therefore, draw his own conclusion as to how far there is a real or fanciful connection between the two curves. In some points there is undoubted similarity; in others, an absolute contrariety. In the rainfall curve, if we take the absolute mathematical definition of a *maximum*—when any value is greater than either the preceding or succeeding ones—there is a *maximum* of rain nearly every other year; but if we consider the broader sweeps of the curve, we may find more resemblance. For instance, the *maxima* about 1805, 1816, 1828, 1837, 1848, 1860, and 1871 agree passably in both curves; on the other hand, the absolutely greatest rainfall in the eighty years was in 1811, a year of *minimum* spotted area; while another very large rain *maximum* also occurred near a time of *minimum* spots in 1841. Then again, the greatest *minimum* but one of rain, in 1870, occurred one year before a *maximum* of spots and only two years before the second largest *maximum* of rain. These latter cases, of course, throw doubt on whether we are justified in finding any periodicity at all in the rainfall curve. Any attempt to smooth or alter these curves by arithmetical or algebraical processes can only lead to illusive results; we must base our opinion of the supposed connection between the two curves on our knowledge of other undoubted irregular periodicities.

In our curve of thermograms, Fig. 60, we pointed out that because there was no obvious trace of diurnal variation of heat on many days, we were not, therefore, justified in saying that there was no such thing as a diurnal heating influence of the sun. All that could be said was that his power had been overridden by more powerful influences. In the same way, the fact that there are heavy rainfalls which have no relation to the extent of sun-spotted area, does not of itself prove that there is no connection between the spots and weather. If we could be certain from any other considerations that there was a real connection between the two phenomena, all that we should be justified in saying was, that whatever influences the spots had on weather, there were other influences which might be much more powerful.

Another great difficulty which we have to face in forming our judgment of the possible connection between the state of the sun and weather, arises from the impossibility of laying down any absolute criterion of what is a rainy year. Rain may be produced by so many different causes, and the difference of amount which is measured in places near one another is so great, that we are left a great deal to our own estimate of values or probability. Thunderstorms are the great disturbance of rainfall statistics. Under those circumstances as much as two inches of rain may fall in one place, and but a few drops in another only a few miles distant. Yearly totals show the same discrepancies.

For example, in the year 1872—a year of sun-spot *maximum*—Buchan has plotted the rainfall within the limited area of Scotland. He finds that while near Aberdeen the rainfall was seventy-five per cent. above the average, the amount at Cape Wrath, about one hundred miles distant, was below the average. Then, of course, the difficulty is, why should we take the returns of one station more than another to compare with sun-spots, when the latter affect the whole world simultaneously?

Rain in the abstract is a mere entity—we must say what kind of rain it was which fell. Was it cyclone rain, or secondary rain, or that associated with thunderstorms? The true criterion of periodicity requires not only an amount of rain which corresponds with the state of the sun's surface, but also rainfall under the same conditions of atmospheric pressure. We must not compare the rain of cyclones with the rain of thunderstorms, unless we can show that they may both be produced by increased intensity of the weather generally. This is, however, sometimes the case.

There is another point which must be remembered in the discussion of questions as to the connection between the sun and the weather. We have shown that weather in the temperate zone is the product of the passage of cyclones, anticyclones, etc., so that we cannot properly talk of the influence of any physical cause on weather in the abstract. We must think how the physical cause would act on the general circulation of the atmosphere. When we discussed the daily influence of the sun on weather, we showed how heat modified in a different manner the ascending currents of a cyclone or the descending ones of an anticyclone.

In the same way, if the condition of the sun's surface does affect weather, the action must take place through the medium of cyclones and anticyclones. We must, in fact, show that in years of sun-spot *maxima* and *minima*, the circulation of the atmosphere is either more intense generally, or that the formation of cyclones, etc., is then in some manner modified.

This view of the true nature of solar action explains some anomalies which the advocates of the sun-spot theory have been unable to explain. They find that in some places the *maxima* of spots are associated with the *minima* of rain. If we try to connect rainfall and sun-spots in the abstract, we are helpless to explain the discrepancy. But if, on the contrary, we realize that an alteration in the solar heat may modify the formation of cyclones, then we can at once explain the apparent contradiction of results. For instance, in the year 1872, to which we have already alluded, the general position of cyclone centres over North-Western Europe was considerably displaced. Instead of lying to the west of Scotland, the centre of cyclone activity appeared to lie between England and Norway. This, of course, made England wetter, and the north-west of Scotland drier than usual; but it will take many years before we are justified in saying that this displacement was due to the influence of solar spots.

It is, no doubt, a very tempting ideal to look at the sun as the prime mover of the atmosphere, and to endeavour to follow variations in the heat or energy of his action into their final products as wind or rain. But when we consider what the real nature of weather is, as revealed to us by means of synoptic charts, we see at once that, though undoubtedly an alteration in the sun's power would sooner or later be reflected in his results, any attempt to deduce one from the other directly must lead to disastrous failure.

Relation to Forecasting.

Though opinions will doubtless differ as to whether we are justified in asserting that there is any connection between sun-spots and weather, there is no uncertainty as to what the value of that knowledge would be to a forecaster.

The author believes himself that there are signs of some real relation between the extent of spots on the sun's surface and the rainfall curve at Rothesay, but how should we fare if we tried to forecast the rainfall for any particular year? The most cursory glance at the two curves of sun-spots and rainfall will show that, if we were to attempt to forecast rainfall on the assumption that the amount would follow the sun-spot curve, we should get just the same unsatisfactory results as if we attempted to forecast the temperature at different hours by reference to the mean diurnal curve of heat. Every meteorological element depends for its value on the balance of several nearly equal forces, so that an attempt to forecast the resulting value by means of the variations of one of these forces can only lead to failure.

So far for the use of the knowledge even of a certain cyclic period in forecasting the character of a year as a whole; and it is still more impossible to use any abstract periodicity in forecasting the weather for any particular day. We shall see in a future chapter that all weather prevision depends on the estimate which an experienced forecaster can make as to the probable path of any cyclone, or as to the formation of a new one. How much would the abstract knowledge that it was a *maximum* or *minimum* sun-spot year help him to form such a judgment? Obviously nothing.

On the whole, then, we may say that though there are certainly very strong grounds for the belief that there is some real connection between the state of the sun's surface and terrestrial weather, still, from the nature of atmospheric circulation, we are unable to utilize this fact in forecasting weather, either for any season or for any day.

CHAPTER XIII.

TYPES AND SPELLS OF WEATHER.

INTRODUCTORY.

IN the foregoing chapters we have devoted our attention more to the nature of the causes which produce weather at any moment than to the sequence of weather for several consecutive days. We have, in fact, rather described the nature of the individual disturbances which form, as it were, the units of weather, than the manner in which these components move or follow one another. The word "weather" is used by meteorologists in a twofold sense. When they talk of the weather at a moment, they use the word in a restricted signification, referring to the appearance of the sky, or to the occurrence of rain, snow, etc. When they talk of weather for a longer period, as, for instance, a wet week, or a cold month, they use the word in a more extended sense, and include the sequence of every meteorological element for the time in question.

We have already mentioned that in the temperate zone the units of weather, such as cyclones or anticyclones, are perpetually moving or altering their shape, and thereby producing changes of weather; or to put it more formally, weather in the temperate zone is the product of the passage of cyclones, anticyclones, or of the minor forms of isobars.

We have also pointed out that all forecasting depends on the limited power which we possess of knowing beforehand what the path of any disturbance is likely to be, or what new changes in the distribution of pressure will probably take place. For instance, if we see a cyclone approaching our own country at eight o'clock in the morning, how can we tell in what direction it will move, or if it is likely to grow more or less intense? If we see an anticyclone, are there any signs by which we can know whether it is going to remain stationary, or to break up and disappear?

When we have examined a very large number of synoptic charts we soon see that, though no two are alike, there is much in common so far as their sequence is concerned. Though a cyclone may move in any direction, and almost with any velocity, nothing is a matter of accident, but certain types of motion are associated with certain types of general pressure distribution.

Our purpose in this chapter is to explain the nature of these changes, by giving in some detail the four great types of weather which occur in Western Europe, with shorter notices of those in the United States and in the tropics. In doing so we will bear in mind the twofold object of all scientific meteorology—the explanation of past weather by reference to the motion of cyclones, etc., and the classification of typical changes with reference to future forecasts.

Long verbal descriptions of complicated weather-maps are not only tedious, but unintelligible to all except those who have made synoptic charts their special study. As our object is to convey an idea of the nature of weather-changes to those who have no previous knowledge of the subject, we shall, therefore, rather trust to copious illustrations of carefully selected specimens, and the reader must look at them and supply his own descriptions. By this means he will learn the character of atmospheric changes and the ways of cyclones by eye, rather than by reference to any written formula. He will see the rapidity with which these changes take place, and acquire that knowledge of the nature of weather which will enable him to form a just conception of the great problems of forecasting.

We shall assume that he has so far mastered the preceding chapters that, when we talk of a cyclone, he knows that it is equivalent to bad weather—warmth in front, cold in rear, wind according to intensity; and that when we say an anticyclone covers any country, that means generally fine weather—always light wind, but blue sky, mist, heat, or cold, according to the circumstances of latitude or season. Also that the direction of wind is given at once by naming an isobaric shape or any portion of it.

Our illustrative charts, mostly on a uniform scale and projection, embrace an area that extends from the Rocky Mountains to Moscow, and from the equator to Greenland. In all, pressures of 29·9 ins. (760 mm.), and all above, are marked by full isobars, while those below are dotted, so that the reader sees at a glance the broad elative distribution of high and low pressure.

Distribution of Pressure over the Globe.

Over the above area the distribution of atmospheric pressure presents certain constant features, namely—

1. An equatorial belt of nearly uniform low pressure.

2. A tropical belt of high pressure rising at intervals into great irregular elevations or anticyclones.

3. A temperate and arctic region of generally low pressure, but in which occasionally areas of high pressure appear for a considerable period.

Weather in the Doldrums.

The equatorial belt constantly covers the Sahara and the Amazon valley, and always narrows over the Atlantic at about 30° west longitude, where it often does not reach higher than 10° north latitude. The shape and depth of this area are tolerably constant.

This is the "doldrums" of the Atlantic navigators. Our charts only show the north side of this area; the south side is formed by an anticyclone, which always lies over the South Atlantic. The doldrums, therefore, form a sort of long hollow, or col, between two anticyclones, and though on the one side the north-east trade blows moderately, and the south-east trade on the other, still in the centre there must be calm. This is well shown in Fig. 68, where we see the symbol of calm between the two trades. These sultry doldrums are much dreaded by sailors, for in them "a ship may lie for weeks on the hot smooth water, under a cloudless sky, with pitch oozing from her decks; a region of unbearable calm, broken occasionally by violent squalls, torrential rain, and fearful lightning and thunder." The general appearance of the sky is a steamy haze, sometimes growing into a uniform gloom, with or without heavy rain; at other times gathering into small ill-defined patches of soft cumulus, or the forms of cloud given in Fig. 13, *a*, *b*, and *c*. After dark there is always a great development of sheet lightning till about two in the morning. As the position of this area only varies very slowly in its annual course a little more north or a little more south, there is nothing to change the weather, which therefore remains of a uniform character.

WEATHER IN THE TRADE-WINDS.

The tropical belt comprises a region of high pressure, rising at variable intervals into great anticyclones. These anticyclones are usually the longest in an east and west direction, and often rise into two or more heads. Their position is generally variable, with the exception of one, which is always found over the central Atlantic. This anticyclone forms a very important factor of the weather both of Western Europe and of the United States, and will be constantly referred to as "the Atlantic anticyclone." The extension south and west of this anticyclone is tolerably constant, while north and east it is variable, sometimes rising as far as 60° north, and stretching over Great Britain and Continental Europe.

The wind blows round this as in all anticyclones. The north-east and east winds on the southern side of the Atlantic anticyclone constitute the celebrated "trade-winds." An inspection of Figs. 68–71 will both show their true nature and correct some popular fallacies as to their position and constancy. It is obvious, from the nature of anticyclone winds, that north or north-east winds must stretch far north on the easterly edge, which accounts for the north-east trade being often picked up off the coast of Portugal. But on the westerly edge of the anticyclone the wind must be more south-east or south, and in practice is lighter and more variable. The centre must be, and is, calm, so that the wind-maps which appear in physical atlases with the north-east trade described as a belt of wind parallel to the equator are most delusive. The degree of constancy in the direction and force of the trades is best gathered from an inspection of the charts. We then see at once that the position of the edges of the anticyclone is perpetually changing, and that the gradients are very variable; so that, as a matter of course, both the direction and strength of the trades vary very considerably.

The weather in the trades is usually fine, and the sky more or less covered with a peculiar small detached cumulus, often called "trade cumulus." This is a small isolated cloud bending backwards from the flat base, as in Fig. 11, *a*, in our chapter on Clouds, which often degenerates into the small lens-shaped mass as in Fig. 13, *e*. Sometimes a thin, hard, broken strato-cumulus covers the sky with such regularity that, when seen in perspective near the horizon, we look at a series of bars, like the leaves of a Venetian blind; but if the gradients are all steep, squalls and showers from cumulus cloud are of frequent occurrence in the trade-wind regions.

Cyclones are rarely, if ever, formed to the south of this Atlantic anticyclone; sometimes, however, they have their origin on its south-west side, when they work round the anticyclone, first towards the north-west, and then towards north-east. These are the West India hurricanes.

The north side of the anticyclone is the birthplace of innumerable cyclones of every size and intensity, which invariably move towards some point of east. These are the cyclone-storms which affect Europe.

Cyclones are also occasionally formed on the south-east side near Madeira; these either work very slowly round the high pressure to the south-west, or else leave the anticyclone and go east over the Straits of Gibraltar. In winter-time, another anticyclone generally lies over Mexico, and the col between this and the Atlantic anti-cyclone forms the most prominent feature in the meteorology of the United States sea-board.

Weather in Temperate Zone.

The temperate and arctic region extends from the tropical high pressure to the pole. Though ordinarily low, the pressure is perpetually fluctuating by reason of the incessant passage of cyclones; yet occasionally persistent areas of high pressure appear in certain portions of it.

As a necessary consequence of this, the weather in this zone must be changeable, with variable winds.

From this brief survey, we see at once the broad features of the climates of the world—the persistent equatorial calms and rains, the regular trades of the tropics, and the variable wind and weather of the temperate zone.

We will now proceed to examine the weather of the temperate zone in some detail.

TYPES OF PRESSURE IN TEMPERATE ZONE.

In spite of the great variability of the temperate zone, there are—with reference to Western Europe—at least four constant types of weather which coincide with four distinct types of pressure-distribution.

1. The southerly, in which an anticyclone lies to the east or south-east of Great Britain, while cyclones coming in from the Atlantic either beat up against it or pass towards north-east.

2. The westerly, in which the tropical belt of anticyclones is found to the south of Great Britain, and the cyclones which are formed in the central Atlantic pass towards east or north-east.

3. The northerly, in which the Atlantic anticyclone stretches far to the west and north-west of Great Britain, roughly covering the Atlantic Ocean. In this case, cyclones spring up on the north or east side, and either work round the anticyclone to the south-east, or leave it and travel rapidly towards the east.

4. The easterly, in which an apparently non-tropical anticyclone (or one disconnected with the tropical high-pressure belt) appears in the north-east of Europe, rarely extending beyond the coast-line, while the Atlantic anticyclone is occasionally totally absent from the Bay of

Biscay. The cyclones then either come in from the Atlantic and pass south-east between the Scandinavian and Atlantic anticyclones, or else, their progress being impeded, they are arrested or deflected by the anticyclone in the north-east of Europe. Sometimes they are formed to the south of the Scandinavian anticyclone, and advance slowly towards the east, or sometimes even towards the west.

These types are so named because the prevailing wind in each is from south, west, north, and east respectively. The connection of these European groups with those of the United States will be considered under the details of each type.

Notice will now be directed to the details of these types—first to their main character and seasonal modifications, together with the indications of intensity, and then to any signs of persistence or change of type when possible.

But however much we study details, the above general view of the distribution of pressure in the earth's surface must never be forgotten, as without that we lose the only clue to the ceaseless and complicated changes with which we have to deal.

SOUTHERLY TYPE.

In this type the Atlantic anticyclone extends very little to the northward; another of the tropical belt of anticyclones covers Mexico or the southern states of the American Union; while a third area of high pressure covers Northern and Eastern Europe.

The North Atlantic is occupied by a persistent area of low pressure in which cyclones are constantly being formed; these beat up against the high European pressure, and either die out or are repelled.

Sometimes, especially in summer, small cyclones arising in the easterly side of the area of depression pass rapidly near the British coasts in a north or north-east direction. In either case it is somewhat rare for the centre of a cyclone to reach the coast-line of Europe, so that generally Great Britain is under the influence of the rim or edge of either a cyclone or anticyclone.

At other times the Atlantic low pressure extends over Great Britain, driving the high pressure eastwards, without forming any definite cyclone. In this case, the indications are for tolerably fine weather and little wind, with a very low barometer—a condition which often excites remark.

This type of weather occurs at all seasons of the year, but it is most common and persistent in winter; in fact, the warmth or otherwise of the winter principally depends on the number of days of this type.

No definite sequence of weather to the United States is connected with the occurrence of this type in Europe. While the Mexican anticyclone is tolerably persistent, cyclones which form in the Hudson's Bay Territory usually pass into the Atlantic and are lost there; but at the same time another totally different class of cyclone forms in the col which lies between the Atlantic and Mexican anticyclones, and moves along the northern edge of the former till they reach Europe. The centres, of course, never touch the American continent, but the gales associated with the western side of these cyclones often do much damage to the United States coast.

The above will be more easily understood by reference to an actual example. In Figs. 64–67, we give charts over a large area, for the four days November 10–13, 1877, at 7.35 a.m. Washington. None of these show the zone of

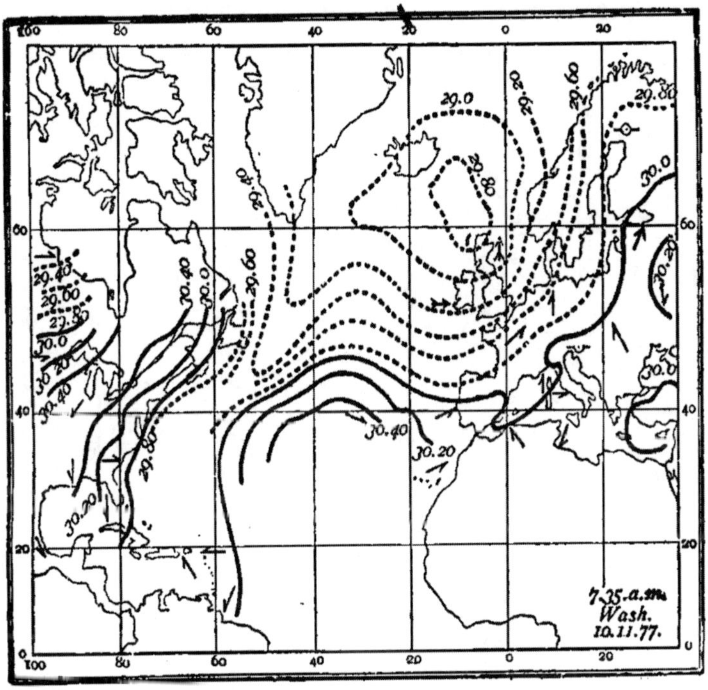

FIG. 64.—Southerly type of weather.

equatorial low pressure, but in all the tropical belt of anticyclones and the temperate and Arctic zone of low pressure are very obvious. In all we find three persistent anticyclones, one over the lower Mississippi valley, another in mid-Atlantic, and a third over Moscow. The North Atlantic and Hudson's Bay Territory are covered by low pressure, and this area is the theatre of the formation of an incessant series of new cyclones, whose history we are now going to trace.

But first let us consider the southern edges of the tropical anticyclones. The east winds under the American high pressure are the trade-winds of Cuba and the Central

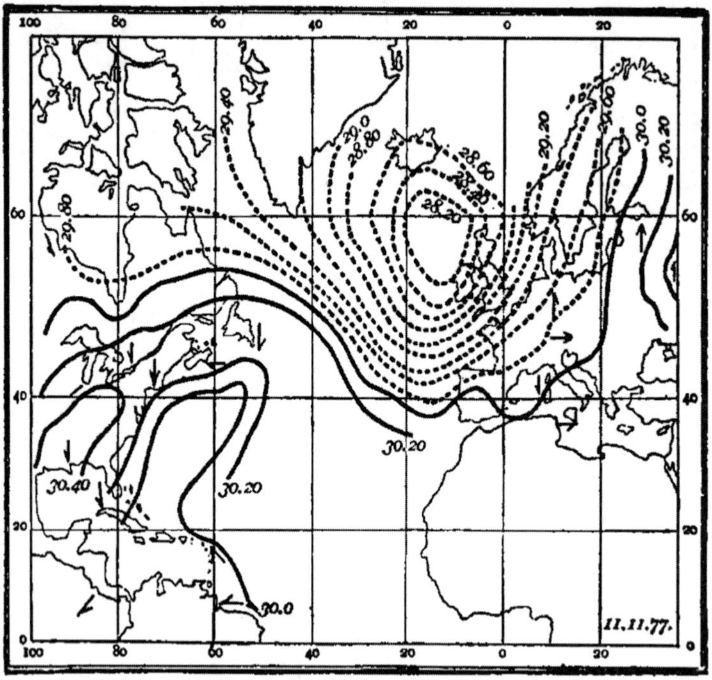

FIG. 65.—Southerly type of weather.

American republics as shown by the small arrows; the Atlantic anticyclone gives the regular trades of that ocean, and the anticyclone, whose edge we see in the chart over Moscow, really extends over the whole of Siberia, and gives the north-east monsoon to the Indian Ocean. This all shows in a very striking manner the dependence of weather in different parts of the world on each other, and also the true nature of the problems which the meteorologist has to solve. The cyclone which covered Great Britain on November 10, 1877, had its origin in the Atlantic anticyclone which dominates the trade-winds. Its eastward path was deflected by the Asiatic

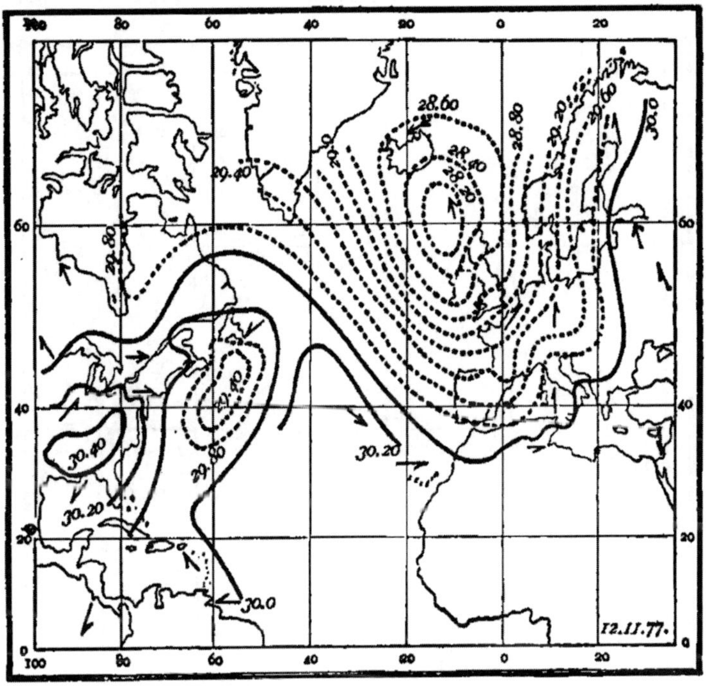

FIG. 66.—Southerly type of weather.

anticyclone which caused the north-east monsoon in Calcutta, and its intensity was increased by a depression which passed into the Atlantic from the Hudson's Bay. At the same time the actual force of the wind was determined at every station by the exposure; every hill drew a little more or less rain, every tidal river brought up local showers. It is this combination of the very large with the very small which constitutes one of the great difficulties of meteorology, and all the skill of the meteorologist is required to assign to each influence its proper place and value. He cannot explain the weather on any day without casting his eyes over the whole

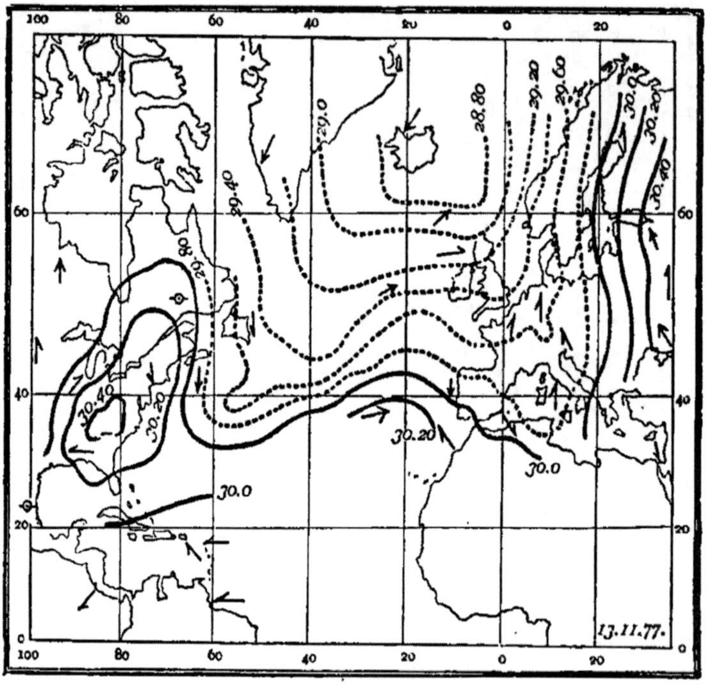

Fig. 67.—Southerly type of weather.

northern hemisphere and round the little hills and valleys which bound his own horizon.

Returning now to our cyclones north of the tropical belt of anticyclones. On November 10 a well-defined cyclone lay between Scotland and Iceland, a V-depression lay in the col between the Atlantic and American anticyclones, while another cyclone covered the Hudson's Bay. Arrows show the general direction of the wind in the leading capitals and cities, and partially the varying velocity.

By next day, the 11th, the position and shape of the European cyclone had scarcely changed, but the depth had increased no less than six-tenths of an inch (15 mm.), while the position of the isobar of 30·0 inch remained the same over Europe. The Atlantic V and the Hudson's Bay cyclone have disappeared and apparently been merged into the great depression which now fills the whole North Atlantic. The nature of this change should be carefully considered, as it is most typical of Atlantic weather, and shows the nature of what the meteorologist has got to deal with, and the impossibility of ever arriving at any calculation as to cyclone paths. If a cyclone would only keep a tolerably regular shape, and move in even a moderately definite path, weather fore-casting would be one of the most certain and definite of sciences. But when, as here, two or three cyclones gather themselves up into a new formation within twenty-four hours, there is nothing definite to trace. We cannot say how the Hudson's Bay cyclone has moved into the Atlantic, even if it is correct to say that it had done so at all. However, such is the way of cyclones, and our object here is to explain it all as best we can. We often see a precisely analogous action when watching the flow of a river. The impulse of two or three small eddies seems to form one big one in a new place.

The effect of these changes on Western Europe would be to cause a rapid fall of the barometer—from surge, not from advance of a cyclone—and to increase the steepness of the gradients with the general intensity of the weather.

The irregular bends in the isobar of 30·0 ins. (763 mm.) over Europe should be noted, for they are due to small secondaries, and indicate rain without wind in their respective districts.

Below the cyclone-region, the Atlantic and American anticyclones are joined by an arm of high pressure, while a very pronounced depression appears over the Bermudas.

On the following day, November 12 (Fig. 66), though the position of the centre and depth of the European cyclone are still unchanged, the area of low pressure has extended over the whole of Europe, which is now covered by a mass of secondaries; and the isobar of 30·0 ins. (763 mm.) has been pushed a little eastwards. Observe that the line of weakness, across which the cyclone endeavours to pass, is the *col* between the Atlantic and Siberian anticyclones.

The loop in the isobars which lay over Bermuda on the previous day has now moved to the north-east and developed into a moderate cyclone, while a third depression appears over Hudson's Bay.

Now look at the last chart (Fig. 67) for November 13, and try to say how it is related to the previous figure. The European cyclone is now represented by an irregular depression over Iceland, whose lowest point is 0·6 in. (15 mm.) above the level of the previous day, but the general sweep of the isobars unquestionably connect this with another depression in mid-Atlantic. The latter certainly represents the cyclone which lay over that region in the preceding chart, much diminished in intensity, and partially coalesced with the Hudson's Bay depression. The European secondaries of the previous day are now represented by a well-marked deflection of the isobars over the Gulf of Lyons. We can describe all this, but how can we trace the history of each individual depression?

While the weather in the Atlantic has diminished in intensity, the low pressure over Southern Europe has extended into Africa; but in spite of all these changes, the position of the isobar of 30·0 ins. (763 mm.) remains very stationary over the eastern shores of the Baltic.

Now, though different totally in detail, these changes are exactly analogous to the fusion of various cyclones into new configurations which occurred in the previous days, and similar changes would continue as long as this type of weather lasted. We might describe the whole roughly by saying that, while the anticyclones remained stationary, the generally low area of the North Atlantic was the theatre of the incessant formation and breaking up of cyclones.

We do not purpose going into the details of the weather-sequence during this type in any one place or country, but the broad features in Western Europe to a solitary observer are very simple.

As atmospheric pressure falls, temperature rises, and the sky grows dirtier till drizzling rain sets in. The wind, from some point of south, having backed slightly, rises in velocity till the barometer has reached its lowest point. As soon as pressure begins to increase, the wind veers a little and gradually falls, the air becomes cooler, and the sky begins to clear; but the clouds rarely become hard, or form well-defined cumulus. By next day, perhaps, the same sequence is repeated, varying only in intensity, but not in general character, and this alteration perpetually lasts for weeks at a time.

The temperature of this type is always high, partly because of the prevailing southerly winds, and, as the cyclones die out, the slight degree of cold which follows is very noticeable. Sometimes a portion of the Russian anticyclone reaches Great Britain, and in winter white frost of short duration would ensue. The air is always damp, principally from the action of southerly winds, and for the same reason the sky is usually dull or overcast. The wind is remarkable for its steadiness, both in direction and the way of blowing. This results from the large scale on which the cyclonic action takes place.

So far for the explanation of weather after it has passed, but we may now consider how this example illustrates the nature of forecasting. Beginning in the west, the United States forecaster has two classes of cyclones to deal with. No rules can be laid down in the abstract by which, given a cyclone to-day, he can calculate where it will be to-morrow. But by experience he knows that the cyclones which form near Bermuda run a totally different course from those which form over Hudson's Bay, and he can generally form a very fair estimate of their probable motion.

In Great Britain it is evident that when a persistent spell of this type is recognized as having set in, the general character of the weather and direction of the wind are at once indicated. The forecaster knows that the cyclones which press in from the Atlantic will never get past, so that his country will always be under the influence of the front only of the depressions. All that is necessary for storm-warnings is to watch for signs of the intensity becoming so great as to give rise to a gale. The example we have just given in Figs. 64 and 65 is very characteristic of a gale coming on entirely from increase of intensity, without any motion of a cyclone. This shows the value of any indications of increasing or decreasing intensity which can be derived from any source.

An inspection of the illustrative charts will show that the area involved is so large that it is hopeless to trace the cyclones as a whole, but that usually within the area of the British telegraphic reports, and always somewhere, there are localities to the east or north-east where the pressure is steady. Over the Atlantic great variations occur, and the forecaster has, therefore, first to try and discover the area of steady pressure, and then to keep a sharp look-out for any rapid fall of the barometer over the west coast of Ireland, which would produce steep gradients and their associated gales. When once a fragment of a ring of steep gradients is formed, its progress eastwards must be traced by telegraph, and watch must be kept that there is no giving way of pressure over Scandinavia. Since the rate of progress of the steep gradients is usually slow and pretty regular, and since, as has been shown above, the direction of the wind with the general character of the weather is subject to little uncertainty, gales of this type are practically forecast with almost greater success than any other class.

The forecaster in Central Europe is not so fortunate.

The nature of the changes there are so complex and so ill defined that he can scarcely follow them after they have happened, so that he can do little more than forecast generally unsettled weather while the barometer is falling and secondaries are forming in sympathy with the great Atlantic cyclones. After the mercury has begun to rise, improving weather is certainly indicated.

The Russian forecaster has a totally different task. He recognizes the type, and knows that as long as his anticyclone lasts there is no fear of bad weather. We have shown that there is always some isobar, in this case 30·0 ins. (763 mm.), which remains nearly stationary, and he has to find this out in each case, and watch for any symptoms of a serious change.

Thus we see, as the foundation of all synoptic forecasting, that the official in charge of the central bureau must learn by experience the ways of cyclones in his own country, and decide each case on its own merits according to the best of his judgment.

The property of any type of weather to continue for any length of time is called the "persistence" of that type. Many phases of weather are due to this principle, and for forecasting it is very important to recognize any signs of this continuance; but, as the indications for this type are the same as for any other, we will describe the details of persistence later on.

Then as to signs of change. This type may merge insensibly either into the westerly on one side, or the easterly on the other, the latter change being usually the more abrupt; but it is not possible to give any detailed description of symptoms of change.

WESTERLY TYPE.

In this type the permanent belt of anticyclones does not extend very far north, and pressure decreases steadily from the tropics towards the north. Under these circumstances, cyclones are developed on the north side of the Atlantic anticyclone, which roll quickly eastwards along the high-pressure belt and usually die out after they have become detached from the Atlantic anticyclone in their eastward course. Their intensity, and consequently the weather they produce, may vary almost indefinitely. When the cyclones are formed so far south that their centres cross Great Britain, and are of moderate size, the intensity is usually great, and severe well-defined storms, with sharp shifts of wind, are experienced. These occur most frequently in spring and autumn, and are the most destructive storms which traverse Great Britain.

In another modification, while the pressure is low to the north, and the isobars run nearly due east and west, the whole of the arctic area of low pressure surges southward, with an exceedingly ill-defined cyclone, bringing a rim of steep gradients along the edge of the Atlantic anticyclone, and across Great Britain, in a manner analogous to the phase of southerly type before explained. The indications then are for rain and westerly gales, with very little shift of wind. This phase belongs almost exclusively to the winter months.

But the commonest modification at every season, and that which forms about seventy per cent. of European weather, is when the intensity is moderate, and the cyclone paths are so far to the north of the British

Islands that the wind merely backs a point or two from the south-west as the cyclone approaches, and veers a point or two towards the west as the cyclone passes, the general direction of the wind being between south-west and west, without rising to the strength of a gale, while rain is moderate in quantity.

Sometimes in summer a prolongation of the Atlantic anticyclone covers the southern portion of Great Britain, and distant cyclones of small energy just influence the northern countries of Europe. Then the intensity is too small to develop rain, and only produces cloud in the middle of the day, so that fine, dry weather is indicated, which when very prolonged may give rise to drought. This is by far the commonest of all weather types in temperate regions, and occurs at every season of the year.

The existence of this type in Europe is sometimes associated with a similar phase of weather in the United States. That is to say, pressure being high over Mexico, cyclones form over the Rocky Mountains, and then pass along the line of the Lakes into the Atlantic. To this class belong almost exclusively the cyclones which pass from the United States, over the Atlantic, into Europe. At other times, a persistent anticyclone may cover the American continent, and the whole of the European system of cyclones is born and developed in mid-Atlantic.

Before we give more details, it may be well to exemplify some of the leading features of this type. In Figs. 68–71 we therefore give charts over the North Atlantic and Europe for the four days, February 26 to March 1, 1865. These may be taken to represent a fair specimen of ordinary broken weather in Europe, without sufficient intensity to give steep gradients and severe gales. In all, the Atlantic anticyclone was flanked on the west by another over the American continent, and

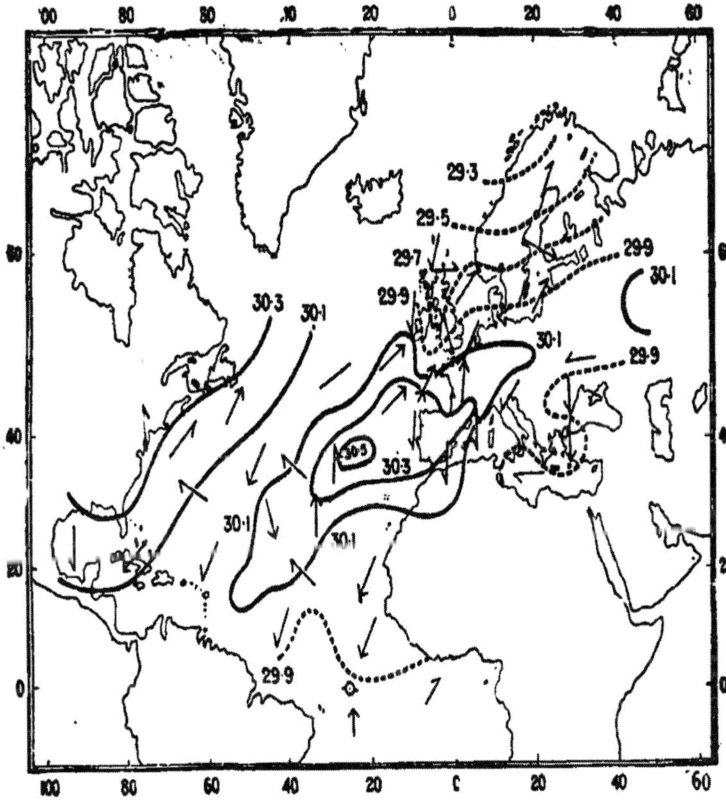

FIG. 68.—Westerly type of weather.

on the east by another over Central Asia. This last only appears in three of the charts. We have therefore to deal with the trade-wind region south of the Atlantic anticyclone; the cols on either side of it; and the slope of decreasing pressure which extends towards the Pole.

We shall take the equatorial region first, because we want to show the nature of weather-changes in that part of the world, but not to have to recur to the subject again till the end of this chapter. On all four days the

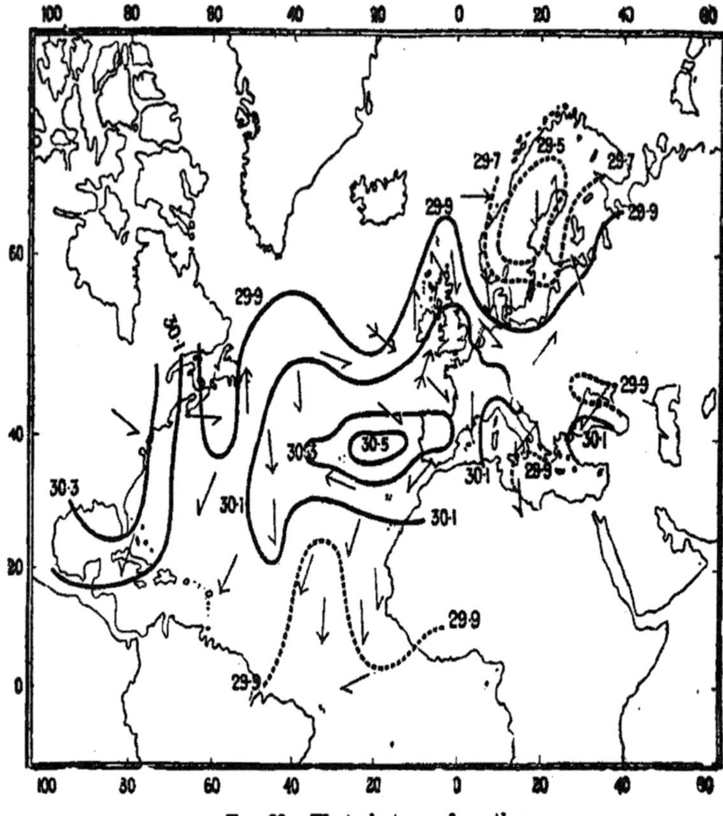

FIG. 69.—Westerly type of weather.

broad features of tropical pressure distribution are the same; that is to say, the type of weather is essentially constant. But in details no two days are alike, for a series of bends in the isobars denote a succession of weather changes, some of which eventually have an influence on Europe. On the first day, February 26 (Fig. 68), the isobar of 29·9 ins. (760 mm.) only shows one bend northwards, while the north-east and south-east trades are separated by a calm near the equator. By

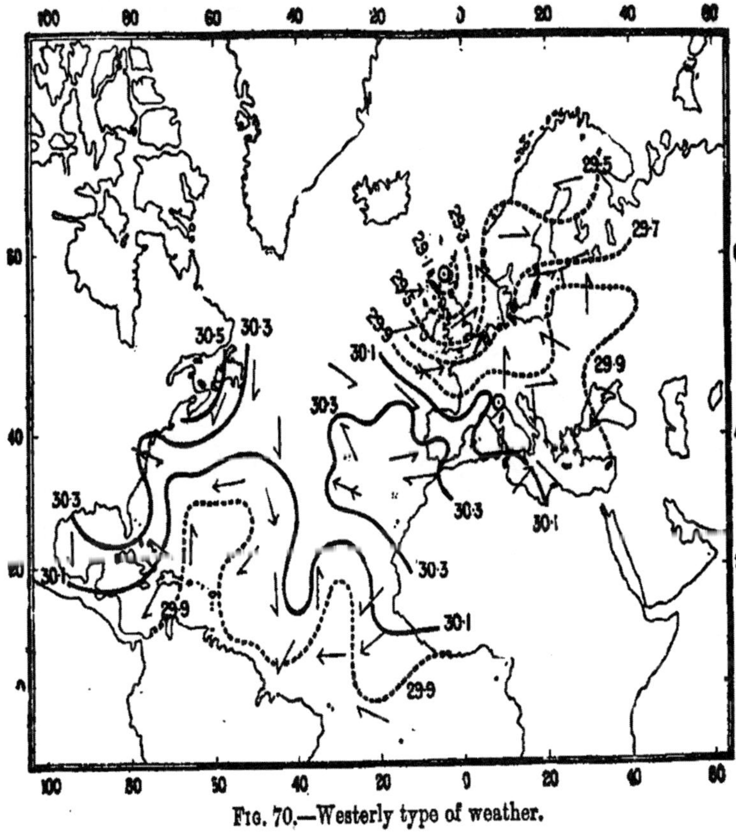

FIG. 70.—Westerly type of weather.

next day this bend had become more pronounced, and moved a little more to the north-east. This latter motion is very interesting, for the prevailing wind is north-easterly, but the direction of the wind has not conformed to the bend of the isobars in the manner which might have been expected.

On the third day, February 28 (Fig. 70), very great changes have occurred. Under the col which lies near Bermuda, a second bend has made its appearance so as to greatly modify the trade-wind region in the West Indies; by next day (Fig. 71) this bend has developed into a well-defined cyclone of very moderate intensity,

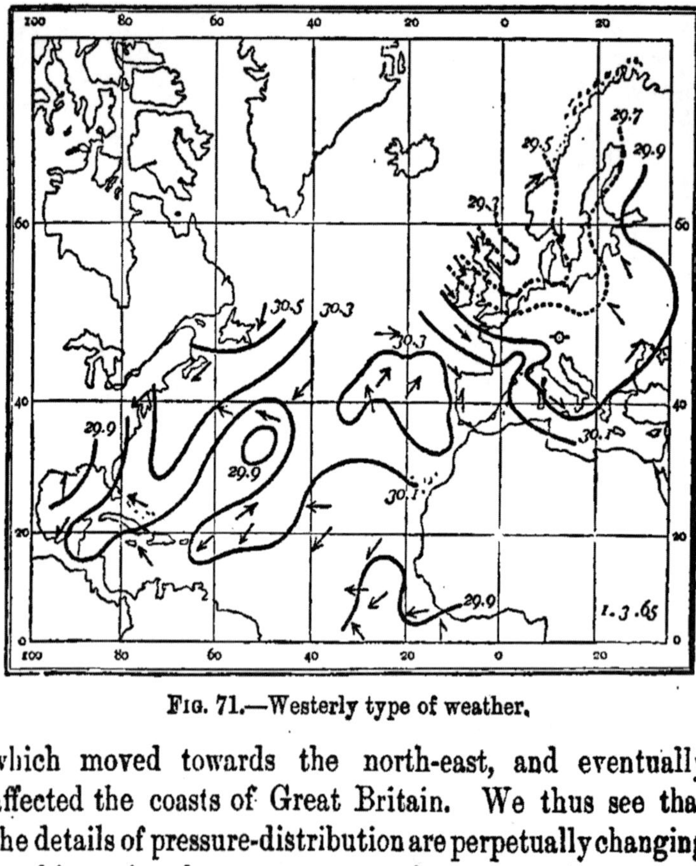

FIG. 71.—Westerly type of weather.

which moved towards the north-east, and eventually affected the coasts of Great Britain. We thus see that the details of pressure-distribution are perpetually changing in this region, but never more than a certain amount. We can therefore easily understand why the weather which is always experienced in these latitudes is described as generally easterly, variable in strength, with the weather fine or showery according to circumstances, but never following the cyclonic sequence of the temperate zone. This modified alternation of weather is called the fluctuation of its type, as opposed to a change of type, which would involve a totally different distribution of pressure.

From this digression on the trade-winds we must now return to the cyclone-traversed region of the temperate zone and the cols of the more tropical parts of the world. On February 26 (Fig. 68), we find a fragment of a large cyclone over Norway, a V over Great Britain, some complex secondaries over the Mediterranean, and an anticyclone over the United States. Note, however, three innocent-looking bends on the north-west edge of the Atlantic anticyclone.

By next day the Norwegian cyclone and the British V have fused or merged into an irregular cyclone which covers Scandinavia; while the Mediterranean secondaries have also formed a new cyclone, and a corner of the Asiatic anticyclone just appears near the Black Sea. Further west, the three bends in the isobars which looked so harmless the preceding day are now reduced to two, but have gained intensity. One lies to the south of Iceland, and the wedge which precedes it determines the weather for this day in Great Britain. The other, which is less intense, lies south of Newfoundland; but the American anticyclone has somewhat retreated.

By next day, February 28 (Fig. 70), the Norwegian cyclone had nearly died out, while the Atlantic cyclone, with its associated wedge, had travelled eastwards and much increased in intensity. In connection with this, a mass of secondaries had developed over Germany and Central Europe in the col which lay between the Atlantic and Asiatic anticyclones. Similar changes are most characteristic of European weather during the persistence of this type, and a knowledge of them is of the utmost importance in forecasting. The cyclone which has come in from the Atlantic is moving and will continue to move towards the north-east, and so far it might be said that it did not affect the forecasters in Central Europe; but when we know that the passage of the depression will develop secondaries and bad weather, it is evident that the indirect influence of the Atlantic cyclone is very great. In every part of the world we may say that the passage of a cyclone in the temperate zone will develop secondaries in the tropical col over which it passes. We may also use this as an illustration of the fact that the tracking of existing cyclones plays but a small part in forecasting, as compared with the larger question of detecting influences which will make new cyclones or destroy old ones.

The col nearly over Bermuda had developed a well-marked inflection near the West Indies.

Lastly, by the morning of March 1 (Fig. 71) all these changes had somewhat developed. The British cyclone had begun to fill up, and the European secondaries had much diminished in intensity. This is an example of what we have already mentioned in the abstract—that a cyclone which is filling up is decreasing in intensity, and *vice versâ*. In mid-Atlantic, the bend in the isobars near Bermuda, as before mentioned, has developed into a small cyclone, which lies between the Atlantic and American anticyclones. The latter has moved a little towards the east.

We may give the general features of British weather for these four days as a sample of the type, and the reader may fill up those in any other country for himself. On the morning of the 26th, the weather in Great Britain was wet and broken from the influence of the V. Next day the weather was beautifully fine from the wedge; the third day, wet and stormy—this time from a true cyclone—and, finally, cold and fine from the rear of the same cyclone on the fourth day. Similar alternations of weather would go on, with endless modifications, so long as the type persisted. From this we see the contrast which the westerly type presents to the southerly one. In the latter, Great Britain was constantly exposed to the influence of the fronts only of cyclones; in the former, both fronts and rears develop their characteristic weathers. We gave the details of the changes of temperature which occurred over the whole of Europe during the first three days in our chapter on Heat and Cold.

In this example the United States was constantly under the influence of a persistent anticyclone, and, so far as it goes, this shows the nature of a type of weather in that country.

We can now easily understand the following particulars of the characteristic weather of this type.

The general temperature of this type is about the average of the season—a little warmer in front of the cyclones, and a little colder in rear. In winter, however, a great prevalence of this type gives an open season, as the high wind prevents frost, unless the cyclones are so far north that the influence of the Atlantic anticyclone is felt.

In summer, on the contrary, if the type be intense, the temperature is below the average, from the excess of cloud hiding the sun.

Another important consideration, as regards temperature, depends on the position of the normal cyclone-path. The difference of temperature just north and south of a cyclone-centre is very marked, so that when cyclones pass further south than usual, the temperature of the region lying between the usual and actual paths is greatly lowered.

To this type also belongs a peculiar class of warm, cloudy anticyclones, which seem to be associated with cyclones passing to the far north, but which have not yet been investigated.

As regards damp, wind, and weather, the most noticeable feature of this type is the changeableness of all these elements. This must be so, because the rapidly moving cyclones bring up alternately the damp, rainy, southerly winds and the dry, cold, northerly currents of their fronts and rears respectively.

The telegraphic forecaster, instead of thinking how cyclones are going to die out, as in the southerly types, has to consider along what paths they will move. No generalities are of much assistance; his opinion must be formed by his own judgment, and from experience of cyclone-paths in his own country. For instance, in Great Britain, he can often tell whether the centre will skirt the north-west coasts of Scotland, or else traverse England on its way to Denmark.

Dr. J. van Bebber has classified the cyclone-paths of Central Europe for the use of the Deutsche Seewarte, while in the United States they know that the great majority of cyclone-paths pass along the line of the Lakes and St. Lawrence valley. But, in spite of any classification, we must never forget that a cyclone may travel in nearly any direction, and for that reason the knowledge of the most usual paths is of less use in forecasting than in explaining the climatic peculiarities of a country.

Northerly Type.

The special feature of this type is the presence of a large anticyclone over Greenland and the arctic portion of the Atlantic, which either joins the Atlantic anticyclone or is only separated from it by a col. On the east side of this, over Europe and Russia, lies a persistent area of low pressure, which is the theatre of the formation of an incessant series of cyclones, while innumerable secondaries are formed over Great Britain and France. The cyclones either move eastwards, or else, if they stand still, surge up and down and alter their shape in a very peculiar manner.

This is, in fact, the exact converse of the southerly type. In that, Europe was persistently under the influence of southerly winds and cyclone-fronts; in this, it is as steadily under the influence of northerly winds and cyclone-rears.

This type occurs chiefly in the winter, spring, and summer; it is very rare in the autumn months.

On the American side of the Atlantic, this distribution of pressure exercises a profound influence on the general character of the weather. Instead of the cyclones finding an easy path into the Atlantic, their eastward progress

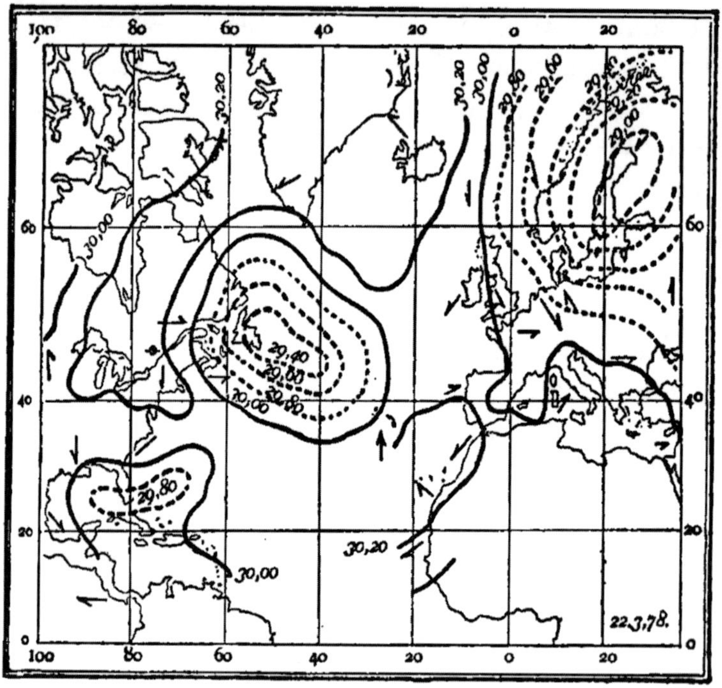

FIG. 72.—Northerly type of weather.

is checked by the areas of high pressure, and in some instances their direction is even reversed.

For instance, in Figs. 72–75 we give reductions from the United States maps of the northern hemisphere, for March 22 to 25, 1878, at 0.43 p.m. Greenwich, or 7.35 a.m. Washington time. In all the Atlantic high pressure will be found stretching far north, till it nearly meets another anticyclone lying over Greenland; and in all, relatively low pressure will be found both over Northern Europe and the western states of the American Union.

On March 22 (Fig. 72), each of these low areas contains a cyclone, one over Finland, giving northerly winds

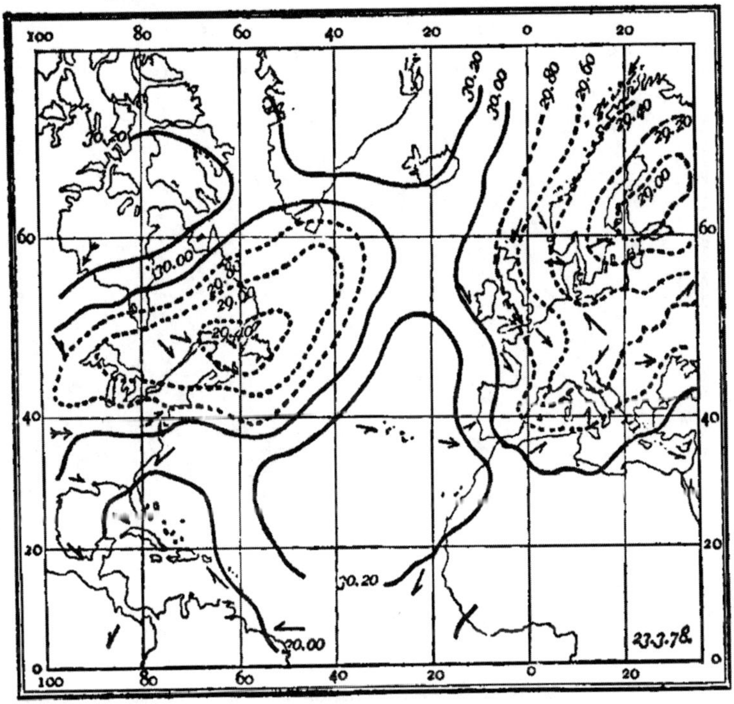

FIG. 73.—Northerly type of weather.

and cloudy weather over Great Britain and the greater part of Europe; the other about three hundred miles west of Newfoundland. An independent cyclone lies near Florida, and a col separates the Atlantic and Greenland anticyclones.

By next day (Fig. 73), though the centre of the

Finland cyclone has hardly changed its position, the area has extended westwards, and the weather over Western Europe becomes rather worse.

Note particularly that the barometer has fallen about three-tenths of an inch in some parts of England, but owing to a surge, and not to the passage of a cyclone.

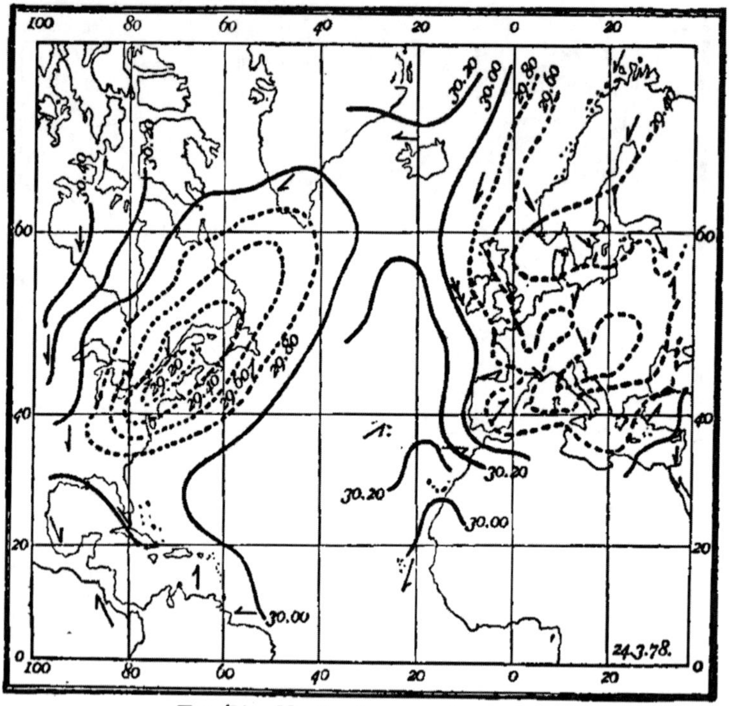

FIG. 74.—Northerly type of weather.

On the other side of the Atlantic, the Newfoundland cyclone has moved westwards, joined the Florida cyclone, and so extended its area as to cover the whole of the northern states. This is the reverse of any we have seen before. The Atlantic anticyclone has enlarged, and projects further north.

By midday of the 24th (Fig. 74), the Finland cyclone has lost any definite shape, while another centre has formed over the Carpathians, and a complicated system of secondaries over Western Europe. The whole is most typical of this kind of weather.

We referred to this chart in our chapter on Squalls,

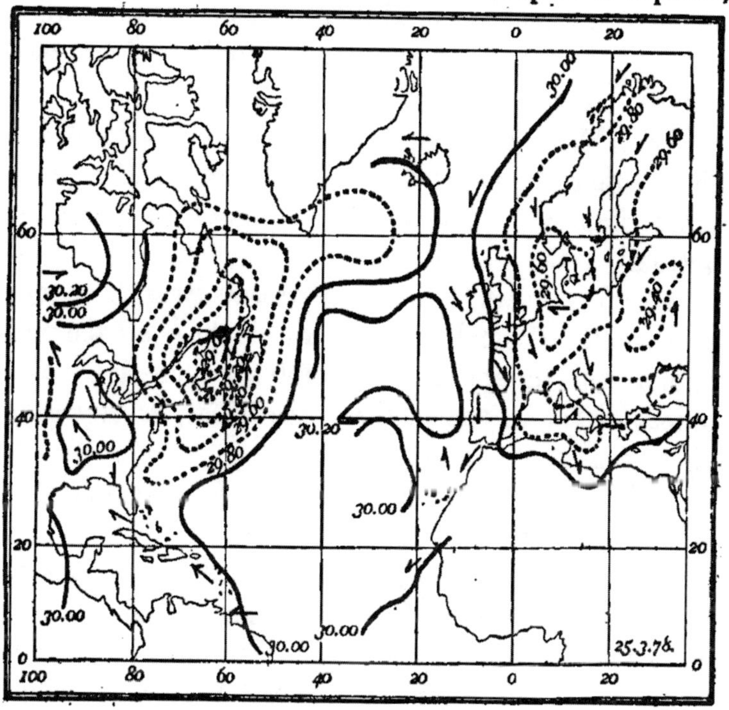

FIG. 75.—Northerly type of weather.

for out of the complex bends in the isobars which we see over England and France developed a V-shaped depression of great intensity, a squall in which capsized the British man-of-war *Eurydice* almost within sight of port.

The American cyclone has moved towards the south-west, and is now centred over the New England states.

It has also slightly diminished in size, but increased in intensity, probably under the action of the anticyclone which lies in the north-west.

Lastly, on the chart for March 25 (Fig. 75), we see that the two centres of the European cyclone have moved as if they were revolving round each other, or round a common centre, while the whole level has risen, and the secondaries have much diminished in complexity.

With these changes, and the rise of the barometer, the weather over Great Britain and Western Europe has much improved, but the wind retains its prevailing northerly set.

Our illustration certainly represents weather-changes of exceptional complexity, but still it shows all the more forcibly the impossibility of applying numerical calculations either to the motions, the winds, or any other phenomena of a cyclone.

This is equally evident when we look on the other side of the Atlantic. The cyclone there has reversed its direction and now gone towards the north-east. Besides this, the intensity has still further increased so as to give worse weather over Canada, New Brunswick, and Nova Scotia, while one secondary projects towards Bermuda, and another in the direction of Iceland.

So long as this type continues the sequence of weather at any station is tolerably simple in Great Britain. As the barometer falls, the wind veers towards the north-east, with a hard, cloudy sky; wind and rain according to the intensity, with an increase of temperature; and then the sky clears, the wind backs by north towards the north-west, and the air gets colder as the mercury begins to rise.

But during the whole continuance of this type, the general northerly set of the wind and the peculiarly hard sky are never lost, and numerous secondaries will give rise to many puzzling contradictions between the movement of the barometer and the severity of the weather.

From this it is manifest that the general temperature of the type must be below the average, and the air must be also dry from the prevalence of northerly winds.

During the persistence of this sequence of weather, all European forecasters have to solve a problem exactly the converse of that which was presented to them by the southerly type. Then they looked westwards for the daily arrival of cyclones, and eastwards for any symptoms of a change of type. Now they look eastwards for a daily formation of new depressions, and westwards for any signs of decreasing pressure over Ireland which would be the forerunner of a different type of atmospheric circulation.

Easterly Type.

In this type the sequence of weather and cyclone-motion turns round the presence of a persistent anti-cyclone over Scandinavia, which profoundly modified the motion of depressions which come in from the Atlantic. The Atlantic anticyclone is, of course, always there; but a col, which is formed between it and the Scandinavian high pressure, crosses Europe and impresses a very definite character on the weather-changes. When cyclones coming in from the Atlantic meet this col, they are either arrested in their course, and remain brooding over the Bay of Biscay, or else they pass through the col in a south-easterly direction. In rare cases cyclones are formed on the southern side of the Scandinavian anti-cyclone, with their centres over Southern Europe or the Mediterranean Sea, and these often move towards some point of west. Nothing can show more clearly than this the value of type-groups in determining the probable course of any cyclone. In the abstract, a cyclone may go in any direction, and in all the European classes we have so far examined they always move towards some point of east; but in this type of pressure-distribution only we may sometimes look for depressions which travel westwards.

This type occurs at all seasons of the year, though it is most frequent in winter and spring, and most rare in autumn. In Great Britain it often persists for two or three weeks consecutively, and gives rise to destructive easterly gales. Nearly one-half of the wrecks on the British coast are due to gales of this class. No direct connection can be traced between the occurrence of this type in Europe and any particular phase of weather in the United States or Canada.

But before we go into details, we may illustrate the nature of this type by an actual example. In Figs. 76–79, we give large charts of a considerable portion of the northern hemisphere for the four days, February 25–28, 1875, at about 8 a.m., Greenwich. In all, an area of high pressure rests over Scandinavia, while the Atlantic anticyclone reaches so far north as to suggest some features of the northerly type. The col of low pressure below these two anticyclones is the theatre of cyclone activity, and we will now describe how the weather in Western Europe was affected by these changes. On the morning of February 25 (Fig. 76), we find the Scandi-

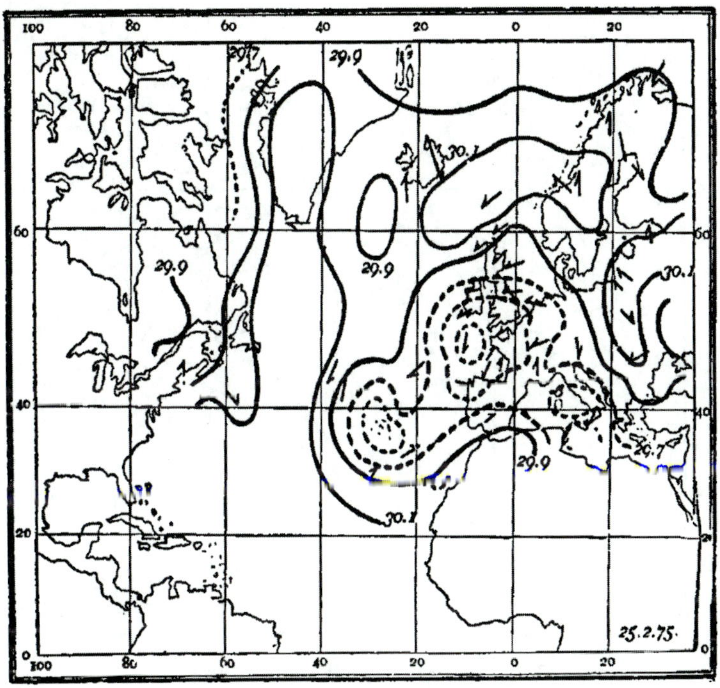

FIG. 76.—Easterly type of weather.

navian anticyclone almost meeting a wedge of high pressure stretching northwards from the Atlantic anticyclone to Greenland. The pressure for several days previous had belonged to the northerly type, with an anticyclone over Greenland, which had now drifted eastwards and joined the Scandinavian anticyclone. To the south of this at least three cyclones are found: one over the Azores, another at the entrance to the English Channel, the third over Italy. These must all be treated as belonging to the same system, as they are all formed in the same pit of low pressure. The weather, of course, is bad all over France, Germany, and Italy. The American

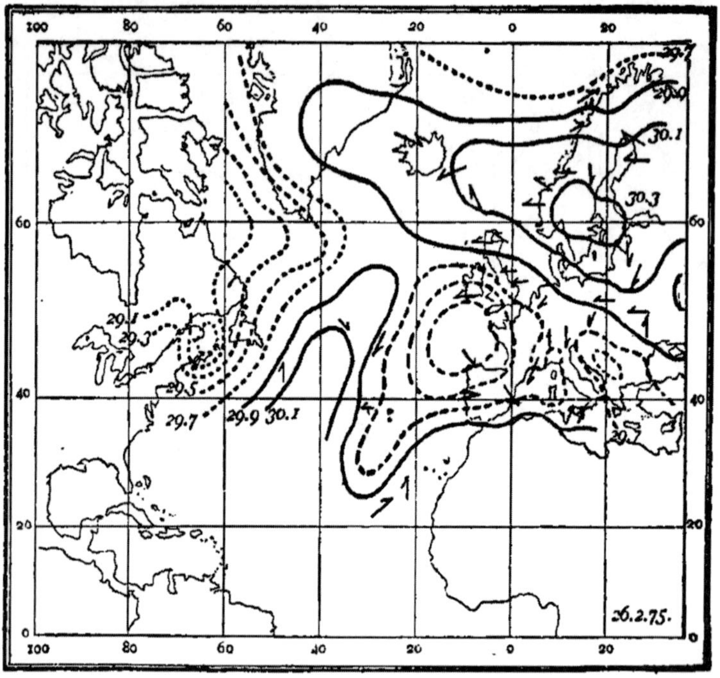

FIG. 77.—Easterly type of weather.

reports are meagre, but point to the existence of a cyclone in Lower Canada.

By next day (Fig. 77), the Scandinavian anticyclone has increased in height, while the Atlantic one has retreated nearly to its usual position. The Italian cyclone has moved a little to the north-east, while that in the Bay of Biscay has apparently moved a very little to the south-west, and so far absorbed the Azores depression that the latter has become degraded into a secondary. Here we have the same fusion of cyclones which we have seen in all the other types, combined with the stationary character which is so peculiar to this class of weather.

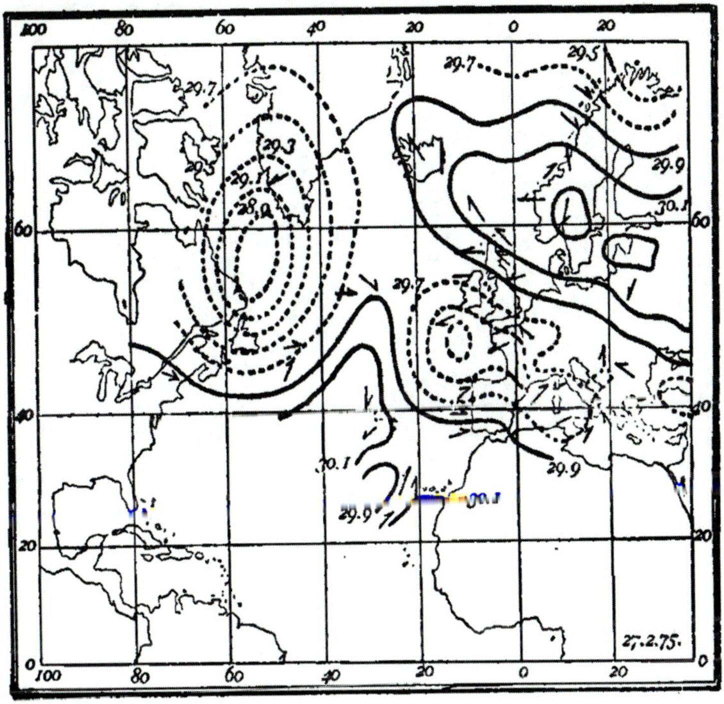

FIG. 78.—Easterly type of weather.

Across the Atlantic an intense secondary has formed over New Brunswick, while another shallow one has pushed itself into the col between the Scandinavian and Atlantic cyclones.

On February 27 (Fig. 78) these changes have made further progress. Though the general position of the

European area of low pressure has not materially altered, the cyclones which lie within it have decreased in complexity, though a new depression has formed in a col between the Azores and the Canaries. The two American secondaries have fused into one large primary, and a large col covers the Central Atlantic.

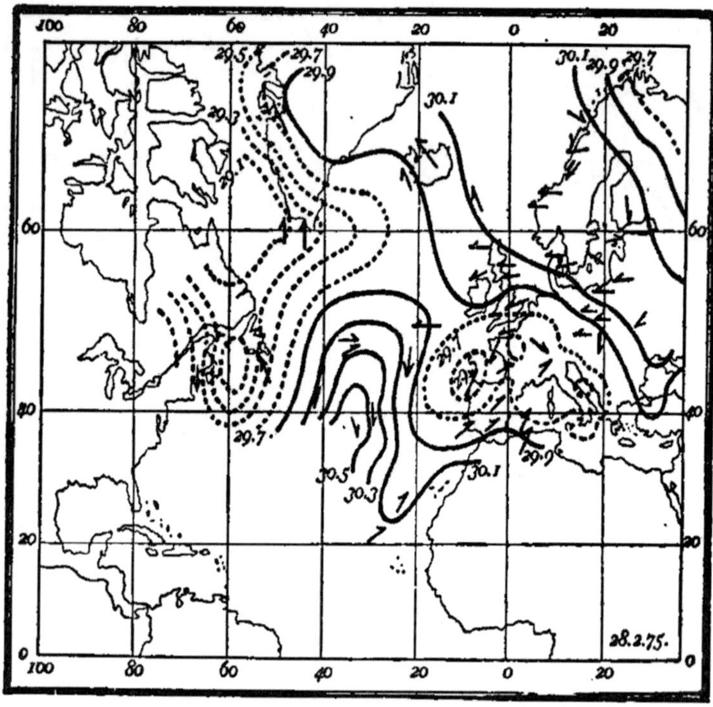

FIG. 79.—Easterly type of weather.

Lastly, on the 28th (Fig. 79), while the Scandinavian anticyclone has diminished in height and area, the Atlantic anticyclone, on the contrary, has increased no less than 0·4 inch (10 mm.) in height, and much increased in size. The size and intensity of the European low pressure has diminished, but its components are more complex; so that while weather has improved over Great Britain, it is worse in many parts of France and Italy. Across the Atlantic, all we can say is that where one large cyclone was yesterday, there are now two secondaries: one intense over Nova Scotia, another slight in the Atlantic col. This is one of the numerous cases where it is impossible to trace the exact history of pressure-changes.

The general character of the weather in Great Britain during the persistence of this type is very well marked. The sky is usually black, and, even if there is a certain amount of blue overhead, the horizon has a peculiar black, misty look, popularly known as an "eastern haze." This is quite different from the misty horizon of a calm day in the westerly type, and is associated with the peculiar bitter feel of an east wind. A well-known saying is—

> "When the wind is in the east,
> It's good for neither man nor beast;"

and this is certainly no exaggeration.

This is the most striking illustration we can have of the general principle that no instrumental records can take the place of verbal description. We might find two north-east winds recorded automatically, of exactly the same velocity and temperature—one on the northern side of a cyclone of the westerly type, the other at the edge of the Scandinavian anticyclone in the easterly type. Read mechanically, they might be taken to be identical, while practically they are very different. Unfortunately we can give no explanation of the malignant nature of true east winds.

The temperature is generally low, but more variable

than during the northerly type. This is because the cyclone-centres sometimes get so far east as to bring up a breath of southerly wind, which is speedily driven back by a new irruption of pressure from Scandinavia.

The wind is always from some point of east, with less tendency to back towards the north than during the continuance of the northerly type, and generally keeping between north-east and south-east. The contrast between this and the westerly type will be strikingly evident if we look back at Figs. 71–73, and note that they refer to the same three days of the year, February 26–28, as our last three (Figs. 77–79). By selecting these dates on different years, all diurnal and seasonal variations are equalized, and the entire difference of wind and weather is solely due to difference of type.

Forecasting during the persistence of this type presents the greatest difficulties, especially in Western Europe. Though the general character of the weather-sequence may be sufficiently obvious, still there is the utmost uncertainty as to the paths of cyclones. When these come in from the Atlantic, we have no means of saying whether they will pass through the col in a south-easterly direction, or whether they will be deflected to a north-easterly course. In addition to this, the motion of cyclones, in whatever direction, is so irregular that the forecaster is doomed to frequent failure.

The signs of persistence are chiefly such as may be derived from watching the position of the Scandinavian anticyclones, and the continuance of low pressure to the west of Ireland. The signs of change, on the contrary, turn round any diminution of pressure in Sweden, or the appearance of high pressure far north in the Atlantic.

The four great types of weather which we have now sketched are capable of being divided more minutely into sub-types; but these would vary so much for different countries, that they cannot be detailed in this work. All that we can do here is to note the universality of the principle, and the properties of weather which the existence of types explains so readily.

Although we have already mentioned the great principles of weather-changes which are designated by the terms "intensity," "fluctuation," "persistence," "recurrence," and "dependence" of type more or less incidentally, it may be well to add a few remarks here on all of them, beginning with—

Intensity.

We have already explained the term "intensity" as applied to single cyclones, and shown both how it is measured by the gradients and how it influences the weather.

But "intensity of type" denotes the character of a sequence of weather to which the epithet of "broken" would be applied. Broken weather is found by synoptic charts to be the product either of small quick-moving cyclones which only exist for a very short time, or of numerous secondaries; in contradistinction to the weather produced by large low-gradient cyclones, moving slowly and lasting for some days, which would be associated with more settled weather.

The relation between these two kinds of intensity is analogous to that between long, single, heavy gusts, and numerous short puffs of wind; both are symptoms of great atmospheric disturbance, though of a different kind in each case.

FLUCTUATION.

The word "fluctuation" is applied to that limited alternation of a general distribution of pressure which occurs every day all over the world. In the tropics, where pressure-distribution is unchanged for months, fluctuation is chiefly confined to small modifications of intensity, which make the weather a little better or a little worse, according to circumstances. In semi-tropical countries fluctuation is much larger, usually from the formation of secondaries, though the general type does not materially change. In the temperate zone, on the contrary, we have not only enormous fluctuation of type, but also complete alteration of the type itself. The classification of phenomena called fluctuation is of the greatest value in handling questions of weather-sequence, as it enables us to separate that which is incidental from that which is essential to any type.

PERSISTENCE.

The word "persistence" describes that prominent feature of remaining pretty stationary which characterizes all pressure-distribution over large areas. This is always concurrent with a persistence of appropriate weather, and in this property of types we find the explanation of many phenomena of weather and of many popular prognostics.

For instance, in Great Britain, an interval of cold weather in winter may be produced by the persistent influence of either the northerly or easterly type; or, if only for two or three days, from the wedge-shaped area of high pressure between two cyclones. So also a drought may be induced either by a persistent anticyclone, or else by cyclone-centres far north, when the intensity is slight, while long-continued rain may accompany almost any persistent type if the gradients be steep.

Then, as to weather-prognostics. It is a well-known saying, that "When grouse come down into the farm-yards it is a sign of snow." The birds are driven down in search of food by the excess of snow already existing on the moors, and so far the prognostic would refer to the past rather than to the future; but, by the principle of persistence, the type which has already given so much snow may be expected to continue for some time, and therefore more snow may be expected.

In Germany there is a proverb, "Fresh snow, fresh cold," which holds good for the same reason.

Similarly, the prognostics, "When a river rises without any rain having fallen, bad weather may be expected," or "Irregular tides are signs of rain," have a significance for the future, though both are caused by past bad weather at a distance; for the persistent type will almost certainly, sooner or later, bring more bad weather over the place of observation.

On the same principle, the prognostic, "Breakers in

shore without wind are a sign of storm," holds on the east coast as well as on the west, but for a different reason.

On the west coast, the breakers have sometimes run on ahead of the cyclone which raised them; but on the east coast this does not occur, as, practically, all cyclones move towards some point of east.

Nevertheless, though the storm which raised the waves has never affected the place where they occur, still it is extremely probable that another of the same series will do so; therefore the prognostic is good, though less certain than on the west coast.

It is also manifest that the principle of persistence has an important bearing on forecasts. Unfortunately, though such types are common, it is not yet possible to define any certain indications of change from one to another. One sign of persistence may, however, be mentioned which rarely fails.

Sometimes a type apparently fails for a day or two, but then is re-established with great intensity. When this occurs, its continuance for a considerable time may safely be predicted. For instance, with the easterly type a small cyclone frequently passes rather far to the east, and the wind shifts to the south-west with increased warmth; but when this dies out the easterly type is re-established in full force. In these cases the appearance of the weather is sometimes very characteristic, for, though the wind is west, the look is that of an east wind, and so obvious is this that the people say "that the east wind is not gone yet."

Recurrence.

We have already explained the tendency of certain kinds of weather to recur about the same date every year so fully in our chapter on Seasonal Variations, that it is unnecessary here to do more than allude to that great principle of meteorology. We shall, however, better understand now how the recurrence of weather is the secondary product of the recurrence of a certain type of pressure-distribution; and that to be a true periodicity of cold, for instance, it is not only cold in the abstract, but cold of the same type which must recur about the same date in most years.

Dependence.

By "dependence" of type or weather is meant the supposed connection between the occurrence of any particular type at one season of the year, and the consequent occurrence of it or of another type at another season.

For instance, there is a common saying in Great Britain, that if easterly winds prevail about the time of the spring equinox, then a great preponderance of easterly winds may be expected during the summer. Put into the language of synoptic charts and types, this means if the easterly type happens to prevail about the 23rd of March, then there will be a tendency of that type to occur more often than usual in the course of the summer.

Again, in most temperate countries, hot summers are popularly supposed to be followed by cold winters, and the latter are thought to depend in some way on the former. This is much more difficult to express in synoptic language, for heat and cold are not always produced by the same causes, and, unless the same type of summer is followed by the same type of winter, the apparent relation of the two seasons is illusory.

The same conception of the dependence of one season on the other is found in the tropics. H. F. Blanford has found that in India there is an apparent dependence or sequence of the summer wet season on the preceding winter rains.

At present we can do little more than note such a relationship of seasons, and cannot say whether there is even such a dependence at all. The older weather-lore seems to have been founded partly on observation, partly on an intuitive belief in the general balance of nature. In the main, the course of nature is constant; if the summer is hotter than usual, a cold winter is required to restore equilibrium, and so on for any other phenomenon of weather.

The middle stage of meteorological investigation seeks to find proof of such relation by comparing statistics of rain at different seasons. Here, of course, the great difficulty is to be certain whether all the rains which we compare are really of the same type.

The latest phase of thought would look for some connection or sequence between the forms and intensities of atmospheric eddies. All we can do is to note the facts for future research; and to remark that at the present time no use can be made of dependence of type in practical weather-forecasting.

Change of Type.

So far we have supposed well-defined specimens of each type, but in practice we meet with many transitional forms. Thus the southerly type may merge by insensible gradations into either the easterly or westerly, but in no case can it grow into the northerly. Similarly the westerly type may approximate on either side towards the southerly or northerly, but never jump suddenly into the easterly. In like manner the northerly and easterly types can only merge into those next to themselves on either side, but never into their opposites. This is obvious when we reflect that the types are determined by the surrounding anticyclones, and that a slight shift of one of these latter may modify the type very materially on either side, while a change to an opposite type would involve a total rearrangement of pressure over the whole northern hemisphere.

In a few cases we have been able to point out signs of an impending change of type, but unfortunately the forecaster is often confronted by very sudden alterations in the whole distribution of pressure over the northern hemisphere. Future research may perhaps some day lead to the detection of more certain symptoms of change, though at present we can say but little.

North-east Monsoon.

But perhaps the nature of European types will be more readily comprehended if we give some illustrations of the Indian monsoons. These will be very valuable both as showing weather-features of a totally different character from any which we have hitherto examined, and as explaining the connection between the fluctuations of weather in the tropics and the more variable changes of the temperate regions. In Figs. 80 and 81 we therefore give isobaric and isothermal charts for January 4 and 5,

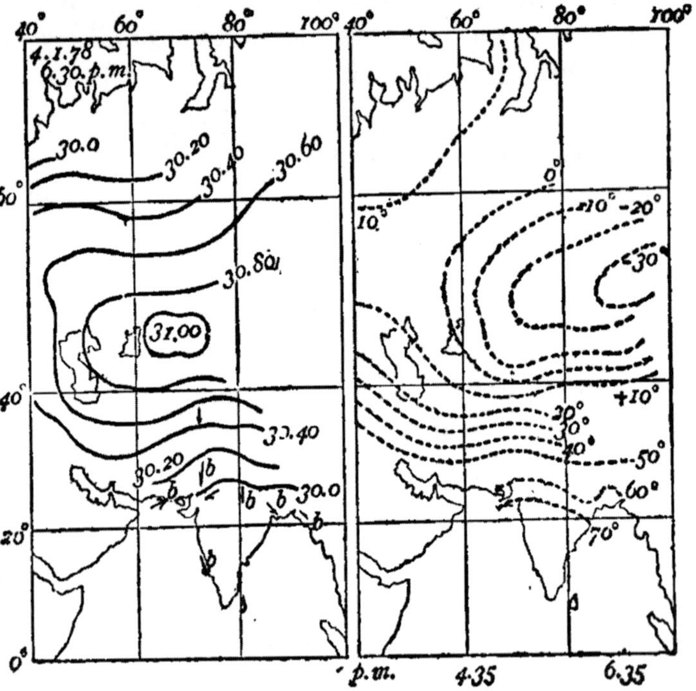

FIG. 80.—North-east monsoon; great cold.

1878, at about 6·30 p.m., Calcutta time—that is, during the season of the north-east monsoon in the Indian Ocean. These charts commence in longitude 40° east of Greenwich, where our Atlantic maps left off, and so continue on the same projection our survey of the world 60° further east.

On both days we find an anticyclone, exceeding 31·0 ins. in height, resting over Tartary, to the east of Lake Aral. North of this, pressure slopes away towards the Arctic Ocean; southwards the pressure falls away to the equator. In fact, this anticyclone is probably the counterpart of the Atlantic anticyclone; while the low

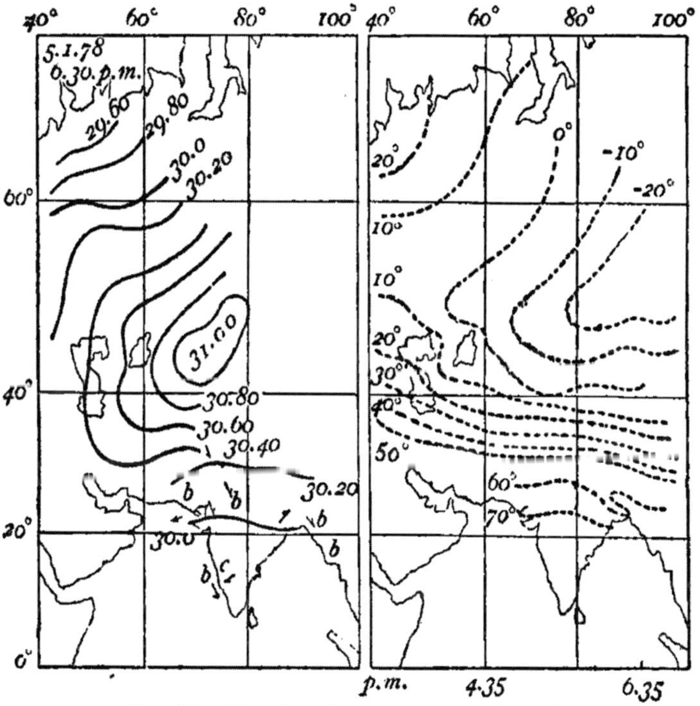

FIG. 81.—North-east monsoon; great cold.

pressure over Southern India corresponds to the trade-wind slope, which we also saw in the Atlantic. The most noticeable feature in both these charts is the persistence of the central Asiatic anticyclone and of the southern slope of low pressure, while the northern slope is more variable; just as we saw in the Atlantic.

The wind circulates round the anticyclone in the usual manner; but note, however, that the winds over Lower Bengal are from north-west, not from north-east, as might have been expected. This is due to a small permanent depression near the mouth of the Ganges, that cannot be shown on so small a map. As this general distribution of pressure lasts all through the winter months, we see that the north-east monsoon is the exact representative of the trade-winds of the Atlantic; only that the result of Asia being a land area is that the easterly winds have a far greater extension northwards than over the ocean.

We have already referred to the temperature-charts in our chapter on Heat and Cold. All that we need do here is to call attention to the great cold—30° Fahr.—near the centre of the anticyclone.

The sky was blue and clear at almost every station in India on both the days in question. What most concerns us here is to note the limited amount of fluctuation over India, and the somewhat irregular nature of the winds, relative to the isobars, in that country. For instance, in Fig. 80 the isobar of 30·0 in. is to the north of Calcutta, while on the following day (Fig. 81) it is a short distance to the south of that city. This, of course, would have been associated with a rise of the barometer, though the general character of the monsoon would not have been affected. This is, in fact, precisely analogous to the fluctuation of the isobars which we saw over the Atlantic to the south of the great permanent anticyclone. Temperature varies in a similar manner, for the isotherm of 60° has been somewhat deflected on the second day by the increasing pressure.

As these charts are very fair specimens of any others during the persistence of this monsoon, we see that the task of the Indian forecaster would be comparatively simple. For, though a limited fluctuation of the general distribution of pressure takes place from day to day, the amount never exceeds a moderate quantity, and still less is the whole character of the weather ever altered in the manner which we have seen in more northern latitudes.

THE SOUTH-WEST MONSOON.

The general character of the south-west monsoon will be best illustrated by giving first a typical sample of two consecutive days, and then by making some remarks on the whole system of Indian weather.

In Figs. 82 and 83 we therefore give charts for June 17 and 18, 1881, over India and Central Asia, at about 6·30 p.m., Calcutta time. These may be considered as typical of the distribution of pressure in those countries during the summer months, just after the burst of the south-west monsoon of the Indian Ocean. In Bengal they relate to the time when the hot season has just begun to give place to the rains.

In both maps we see an oval isobar enclosing pressure less than 29·4 ins. round Lahore, in the Punjab; and in both an isotherm of 100° Fahr. (38° C.), nearly conterminous with that isobar. This depression is usually more distorted by secondaries than on these two days.

The winds blow, on the whole, round the lowest pressure in the usual manner, being from west or south-west to south, and from north-east or east on the northern side of the low pressure. The few wind-arrows which our diagrams admit of will, however, show that the relation of wind-direction to isobars is not so constant as in higher latitudes. For instance, the winds in Lower Bengal are more from the north-west than the general laws of wind-rotation would have indicated.

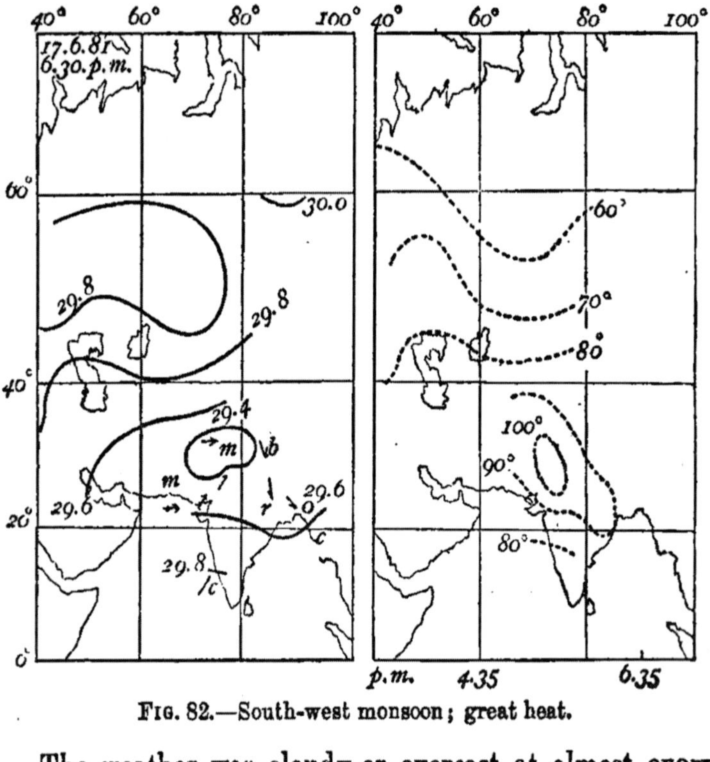

FIG. 82.—South-west monsoon; great heat.

The weather was cloudy or overcast at almost every station, with rain at several, and blue sky at only one on either day; but all the rain is either of the secondary or of the non-isobaric type, and cannot be located by looking at the charts.

It is very interesting to contrast this weather with

that of the north-east monsoon. In the latter there is a difference of one inch of pressure and 100° of temperature between Central Asia and India; in the former only six-tenths of an inch of pressure and 40° of temperature. In the cold monsoon there is scarcely anything but blue sky; while with the warm south-west wind the heavens are almost entirely covered by clouds.

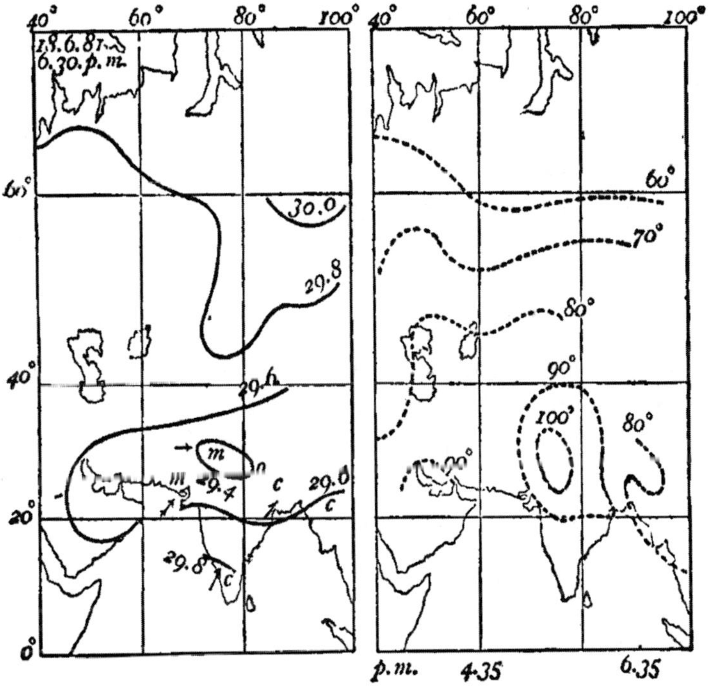

FIG. 83.—South-west monsoon; great heat.

An important but difficult question will immediately present itself as to the relation which the low summer pressure over India bears to the equatorial belt of low pressure over the rest of the world.

In these last two maps we no longer find any arrangement similar to that which occurs in the Atlantic. The persistent pit of low pressure over Scinde and the North-Western Provinces of British India is probably the representative of the equatorial belt of low pressure which constantly covers the Atlantic about 10° north latitude. Here, however, the lowest point is nearly in latitude 30° north; but we know that there is no lower pressure between it and the equator.

The greatest difference is in the absence of an anti-cyclone north of the equatorial low-pressure belt. In Fig. 82 there is a very narrow arm of high pressure between the two isobars of 28·8, and just the fragment of an anticyclone stretching over from the north-east of Siberia. This latter is very persistent over that region at this season of the year, but the difficult point is to determine the relation of the Indian low pressure to the low pressure which usually lies over European Russia in the summer months. This we are unable to give. But we may notice that a somewhat similar phenomenon appears on a smaller scale over Mexico in the summer. Then the Atlantic anticyclone, and another one in the Pacific, about the same latitude, form a col over Mexico; pressure at the same time is persistently low over the United States and also over Central America. Then the distribution of pressure over the whole American continent has some analogies to that over Asia at the same season of the year.

Many meteorologists have contended that this circular persistent depression over Upper India should be considered as a stationary cyclone; but the author's researches have conclusively proved that if we take cloud-forms as a test, none of the true monsoon rains partake of the character of a primary cyclone-front—most of the rain falls from cumuloform clouds—while that in front of a Bengal or any other cyclone seems to grow out of the air. This Indian depression seems to be somewhat analogous to the pit of low pressure which covers the North Atlantic during the winter months. Neither are cyclonic, but both are the theatre of atmospheric disturbance. The former only breeds thunderstorms and secondaries; the latter, well-developed primary cyclones.

We have already described, in our chapter on Non-isobaric Rains, the remarkable character and unknown origin of the rain in this south-west monsoon; and how, without any marked change in the shape or position of the isobars, the dry, hot south-west breezes are suddenly converted into wet and stormy winds. There can be little doubt that the source of this change is to be found in the upper currents which feed the south-west surface-winds, but the subject is too obscure to be discussed in this work.

What mostly concerns us here is the nature of the limited fluctuation of isobars and weather over India; for this is typical of the origin of the modified day to day weather-changes in the tropics as opposed to extreme changes of the temperate zone. There is little fluctuation in the two charts we have given in Figs. 82 and 83; but the bend in the isobar of 29·6 below Calcutta is less pronounced on the second than on the first day. This would be associated with a slight improvement in the weather on the second day.

Sometimes, however, larger changes take place during the continuance of this monsoon. Once or twice during the rainy season—June to October—the pit of low pressure near Lahore stretches further east, and becomes less pronounced, and less distorted by secondaries. H. F. Blanford has shown that this fluctuation is associated with that period of drier weather which is known as a "break in the rains."

We may correct here a popular error that during the rainy season in any part of the tropics it rains all or even every day. Sometimes, no doubt, rain may fall twenty-nine out of thirty days, or for forty-eight hours without cessation; but there are always periods of less intensity and of more intermittent showers.

Returning to India in particular, as the season gets on, small secondary cyclones occasionally form over the Bay of Bengal, and, advancing nearly northwards, strike land on the Orissa coast, and continue their course undisturbed by mountain or valley till they reach the great chain of the Himalayas. In this these secondaries contrast in a marked degree with the great primary cyclones which form on the Bay of Bengal at the change of the monsoon. These latter almost invariably break up when the centre reaches the land. The main characteristic of the secondaries is light wind with torrential rain; as much as fifteen inches of rain has been collected within twenty-four hours during the passage of one of these small depressions.

Occasionally, during the month of May, primary cyclones of considerable intensity develop over the Bay of Bengal, and move towards some point of the west, usually to north-west; and again in October, as the south-west monsoon gives way to that from the north-east, cyclones of very great intensity form in the same Bay, and these too are propagated towards the west or north-west. In either case the coasts exposed to their influence experience very bad weather, with rain and wind of hurricane force in the October cyclones.

From this very brief sketch of monsoon weather we may, however, learn that the persistent seasonal weather in the tropics is exactly analogous to the persistent types of weather which appear periodically in temperate regions. Both are primarily caused by the distribution of pressure, but the changes which only occur once a year in the tropics take place at short and at irregular intervals in higher latitudes, while between the two extremes we find an intermediate series of recurrent types which are neither so regular as in the tropics nor so uncertain as in Western Europe. In both the differences of weather from day to day are due to the fluctuation of the type, or to small alterations either in the shape or intensity of the depressions which form in the low-pressure areas.

We also see problems of forecasting totally different from any which present themselves in the European or American offices, and had we been able to give a greater number of examples, we should have found the truth of the general principle that no mechanical rules can be laid down as to the probable path of a cyclone, or the fluctuation of a type, but that the forecaster who knows the ways of the barometric movements in his own country can usually form a very good opinion as to their future progress.

It may be advantageous here to pause a moment and survey the general aspect of the great problem of weather as it is now presented to us. We have seen over a large portion of the world the same seven forms of isobars constantly reproduced, though with details greatly modified, not only by the type, but still further by the size and intensity, by the time of day, the season of the year, and also by local causes.

But though these sources of variation prevent our writing down on a chart more than the general character of the weather under any isobars, their classification, grouping, and co-ordination each in its proper place, enable us to distinguish the essential from the more accidental features of weather.

And when we come to watch the ceaseless changes of isobars, we see that sometimes cyclones disappear so quickly that within twenty-four hours no trace of their existence is to be found, while at other times the same cyclone may exist for weeks together. Then also we see that a cyclone may either remain stationary, or move in almost any direction with a very wide range of velocity; and we learn the still more curious phenomenon of that fusion of one or more cyclones into a single system which so often makes it impossible to track the paths of depressions.

We also see very clearly how the old idea that a cyclone is necessarily a destructive storm is no longer tenable; and that we must adapt ourselves to the conception that a cyclone is an eddy of very variable intensity, always rainy and always surrounded by a very definite rotation of air, though the force of the wind may vary from a zephyr to a hurricane. When we talk of cyclonic weather we must use descriptive epithets such as moderate, intense, etc., to denote the general force of the wind.

It is also manifest from the great scale on which changes of pressure-distribution take place, that there is some greater cause at work behind them than any local developments of heat or rain. This cause is undoubtedly the general circulation of the atmosphere from the hot equator to the cold Poles; though doubtless temperature and precipitation have a modifying effect on the greater changes. If the earth were surrounded by a vapourless atmosphere, cyclones and anticyclones would undoubtedly be formed, though not the same as those with which we are so familiar.

Now that we know what weather is, we may consider how far it can be forecast more or less in advance.

CHAPTER XIV.

FORECASTING FOR SOLITARY OBSERVERS.

Nature of the Problem.

A comprehensive view of weather-science divides itself into three problems—one direct and two inverse. The direct problem of weather is to explain by mechanical causes the origin and nature of all the complicated phenomena of wind and weather which present themselves to our senses, and the nature of the sequence of weather-changes. This we have already partially done in the preceding pages. The inverse problem of meteorology is, given a portion of a sequence of the weather, to tell what is going to follow. The morning is fine, but now cirrus begins to form, and the mercury has begun to fall—what weather is coming? Last night a cyclone lay over Ireland, this morning it covers Wales—what will the weather be over Great Britain for the rest of the day?

These two illustrations point at once to a natural subdivision of the questions of forecasting: the best that a single observer can do, who has his eyes to look at the appearance of the sky and any instruments at his disposal; and the best that a meteorologist can do, who is seated in a central *bureau*, with abundant telegraphic intelligence for many miles round the country for which he has to issue forecasts, so as to enable him to construct synoptic charts at such intervals as he may think necessary. The latter doubtless represents the highest development of which forecasting is capable; but the former can never be superseded for use among sailors, fishermen, and shepherds. For this reason we will discuss them in separate chapters, and we will take the problem of a solitary observer first, as it is the older and the more generally useful. We shall only attempt to give general principles, and not to go into all the details for any one country.

Prognostics.

We have already gone very fully into the subject of prognostics, and pointed out both the reasons for their success as well as for their failure. When we come to look at all that has been done, we see that, on the whole, we have not been able to develop the practical utility of prognostics very materially, though we have been able to place the whole branch of the subject on a scientific basis.

The most valuable addition of recent times to weather-lore is undoubtedly in the methodical observation of cirrus clouds. The recognition of cirrus as a sign of rain is as old as meteorology, but the deductions which can be made from the direction of the motion of the upper clouds are quite of modern date. No absolute test can be given for the discrimination of fine weather from dangerous cirrus beyond the general surroundings and experience of the observer; but Ley has shown the importance of noting by eye the velocity of the cirrus, because rapid-moving cirrus is a much worse sign of the weather than slowly moving cloud. This is probably one of the most important advances which has been made.

The Barometer.

We propose rather in this chapter to deal with the value of the indications which the barometer can afford to a solitary observer, and especially to explain why the indications of that instrument so often fail.

Why do we sometimes have rain with a rising or steady barometer, and why is the weather sometimes fine with a falling barometer? Then, again, why do we sometimes experience a heavy gale with only a slight fall of the mercury, while at other times the barometer will fall very low without any unusual amount of wind?

These apparent anomalies in the indications of the barometer occur all over the world, and those in each country must be explained by reference to the meteorology of the place. Though we shall draw our illustrations from Great Britain only, the principles which we shall lay down are of universal application. In no branch of the subject shall we find synoptic charts more indispensable, for without them no explanation could ever have been afforded of irregular barometric fluctuations.

General Indications.

The preceding chapters will have sufficiently explained the reasons for what we may call the generally correct indications of the barometer. We can now readily understand why the rapid rise in rear of a cyclone indicates unsettled weather, and the gradual rise of an incipient anticyclone settled fine weather; also why the steady barometer of a persistent anticyclone indicates dry seasonable weather, and the rapid fall of an oncoming cyclone presages storm and rain. All these indications of the barometer can be detected by intermittent observations, or, in fact, by merely looking occasionally at the instrument.

The author has, however, discovered that we can sometimes utilize the greater refinements of self-registered barographs to deduce some knowledge of the future force of the wind from flexures in the recorded curves. These deductions are of the more value now that efficient barographs are so cheap as to be within the reach of everybody.

Author's Rules for inferring from a Barogram whether a Gale is going to increase or decrease.

The principle on which the author's rules are founded depend on what is called the "direction of curvature" of a curve. In the lower portion of Fig. 84, the portion of the trace near the letter A has its hollow turned upwards, and is called convex, relative to the base line. A little further down, near the figures 14, the curve is hollowed downwards, and would be called concave. The other half of the curve is convex almost throughout. From this we see that both convexity and concavity are independent of

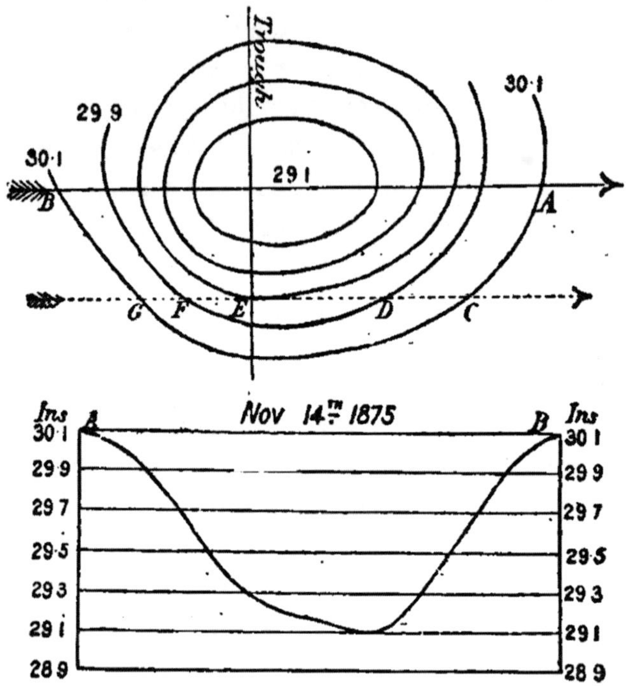

FIG. 84.—Gradients, and flexure of barogram.

whether the mercury is rising or falling, and also of the rapidity of their rise or fall.

If the barometer change at a uniform rate, either upwards or downwards, it is evident that the resulting trace will be a straight line, either rising or falling, and it does not the least matter how rapid the rise or fall is. If, however, the rate of fall changes with diminishing pressure, then the curve will become convex or concave, according as the rate increases or decreases. For instance, suppose, as in Fig. 85, F, that the barometer fell two-tenths of an inch between one and two o'clock, and another two-tenths between two and three o'clock, the resulting barographic trace would be a straight descending line, like *s*; if in the second hour the mercury fell three-tenths of an inch instead of only two-tenths, the resulting trace would be a convex, like *x*; while if it only fell

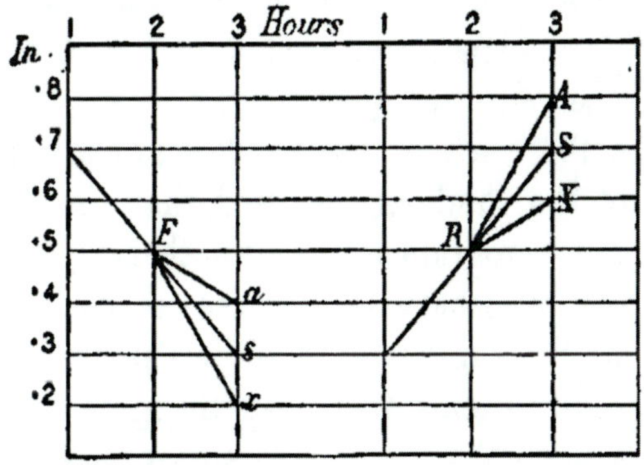

FIG. 85.—Illustrating the origin of convex and concave barograms.

one-tenth in the second hour, the trace would be concave, as *a*.

If we define the barometric rate as the number of hundredths of an inch which the mercury moves, either up or down, per hour, the above may be put in this form.

With a falling barometer, the trace is convex for an increasing rate, concave for a decreasing one. A glance at Fig. 85, R, will show that for a rising barometer the converse is the case; for when the rise is greater the second than the first hour, the trace is concave, as in A; but when less, then convex, as at X; and this result may be stated as follows. With a rising barometer, the trace is convex for a decreasing rate, concave for an increasing one. This is the reverse of what happens with a falling barometer. Now, the simplest and commonest case of barometric change occurs when the centre of a cyclone drifts past a station; the fall of the barometer is then proportional to the steepness of the gradients. When steeper gradients approach, the barogram will become convex; when slighter gradients arrive, the curve will be concave. The converse holds good for a rising barometer: when steeper gradients approach, the curve is concave; when slighter, then convex.

Now, as the force of the wind is proportional to the steepness of the gradients, we find that the direction of curvature of a barogram tells us whether a gale is going to get worse or otherwise, because we can tell if the gradients are becoming steeper or otherwise. We must be very careful to remember that, though a rapid rate of fall is in a general way a worse sign of weather than a moderate one, the indications deduced from the curvature of a barographic trace depend on the variation of the rate, and not on the rate itself. For instance, in Fig. 85, the top part of which gives the isobars over Great Britain on November 14, 1875, at 8 a.m., the crossed line denotes the direction of the cyclone, and an unsymmetrical arrangement of the steepest gradients with reference to the centre is very obvious.

To get the barographic section of a cyclone, or to find out what curve the propagation of the depression would leave on a recording instrument, we have to draw a line across any portion of the plan, as shown on a synoptic chart, parallel to the path of the cyclone, and then, by measuring the distance in time between any two consecutive isobars, we arrive at the flexures of the trace. For the sake of simplicity, we will suppose, in the first instance, that we are stationed exactly on the line of the path of the cyclone, so that the centre will pass over us. By this we make the line of section of the cyclone coincide with the line of gradients, which is not the case in any other portion of the depression.

In the lower part of Fig. 85 we give such a section of the cyclone, sketched in the upper portion, along the line A B. The position of A and B are obverted in the section, so as to read from left to right like an ordinary barogram. Then we see that as the cyclone approached the gradients got steeper, so that the rate of barometric fall increased, and therefore the trace was convex; during this period the gale got worse. After a time, as the ring of steep gradients passed, and the slighter gradients in front of the centre approached, the rate of fall of the mercury decreased and the trace became concave, though still going downwards. The gale moderated somewhat during this time. The passage of the centre marked the turn of the barometer; but as the distance between each consecutive isobar increased regularly after 29·5 inch, the resulting barogram was convex after that level. The actual curve for the day, as given at Stonyhurst, which lay almost in the line of the centre, differs only slightly from this. Thus we see that the normal barographic trace in a cyclone is simply the reflection of the typical shape of isobars in that kind of depression, and that, moreover, to a single observer the direction of curvature—that is, the convexity or concavity of a barogram—enables him to tell whether more or less steep gradients are approaching, and therefore whether a gale is going to get better or worse. There is, however, one limitation which considerably detracts from the value of this deduction. If the line of section of the cyclone which passes over the observer is not square to the isobars, the relative distance between any two consecutive isobars is no longer a measure of the gradients. For instance, if the cyclone in Fig. 84 had passed over an observer anywhere on the line C G, his trace from C to E would have been concave, because C D is a shorter line than D E. But all the time he is getting into a region of steeper gradients, as measured square to the isobars, and therefore the criterion of increased gradients fails. But if a concave need not be an absolute test of decreasing gradients, a convex trace can never fail to indicate steeper gradients with a falling barometer. This may be readily seen by considering the nature of concentric lines.

Conversely, with a rising barometer, we see, in Fig. 84, that from E to G the barogram will be concave, though the gradients are decreasing; but under no possible conditions could a convex trace fail to indicate a decreasing gradient. The author's rule is, then, as follows:—Assuming that the force of a gale is proportional to the gradients, a convex barogram is always bad with a falling, and good with a rising barometer; a concave trace is sometimes a good sign with a falling, and not always a bad indication with a rising barometer.

This rule, of course, involves the supposition that the motion of the barometer is solely due to the propagation of isobars over the observer, but in practice much more complicated changes sometimes occur.

For instance, in a very common class of gale belonging to what we have described as the southerly type of weather, a cyclone, after arriving near the British coasts, remains stationary, but increases, maybe, half an inch in depth. The fall of the barometer which then occurs at any station is no longer of the same kind as that which we have just examined, and the flexure of the trace is determined by other considerations. The direction of curvature would then depend on any variation of the rate of deepening, not on the motion of the cyclone.

For instance, suppose a stationary cyclone which began to deepen from increasing intensity—if the rate of deepening was constant, the trace would be a straight descending line; if the rate increased, the curve would be convex; if it decreased, concave.

But, as we know that the deepening of a cyclone means increased intensity, we may look on a decrease of that rate as a favourable sign, and therefore the indications of the relation of curvature to weather would remain good. The complications which arise from a deepening or shallowi g moving cyclone need not be discussed here, but it is important to notice the two distinct causes of barometric change—the passage of a moving cyclone, and the deepening of a stationary one.

Apparent Failures of the Barometer.

So far we have dealt with what may be called the regular movements of the barometer, that is to say, movements which are associated with or followed by the weather which was anticipated. But we must now explain certain cases in which the weather and the barometer do not seem to be connected in the ordinary manner, and show how, in spite of apparent anomalies, the same general principles of meteorology hold throughout.

CIRRUS BEFORE THE BAROMETER.

In the regular course of isobaric movements, there is one case in which cloud, and sometimes rain, forms before the barometer begins to fall, though almost immediately the mercury turns downwards and falls fast with increasing rain. This happens just in front of the crest of a wedge, and it is for this reason that in the diagram of wind and weather in wedges which we gave in Fig. 7, we placed the word "halo" partly in front of the line of the crest. This is quite common in Great Britain, and it often causes comment that the cirrus begins to form before the barometer indicates the approach of rain. Here, in fact, the sky speaks first, but not so soon as the isobars. If any morning a British forecaster saw a wedge lying over Ireland, and blue sky was reported from the east of England, he could safely forecast that cirrus would appear in the course of the day, before the barometer began to fall.

On the other hand, the author has discovered that, in the tropics especially, the ominous sunsets which precede a hurricane are developed often twenty-four hours before any appreciable depression is formed anywhere; and of course squalls and secondaries have threatening skies as their forerunners without any definite barometric indications. We may lay it down as a general rule that when the sky threatens while the barometer says nothing, something bad is coming; but whether thunder, squall, or gale depends on circumstances.

Rain with a Rising Barometer and an East Wind.

Rain with a rising barometer and an east wind is so common in England that Admiral Fitzroy engraved it on the scales of his barometers as an exception to the general rule that the mercury fell for rain. No explanation was, however, attempted, and, in fact, could not have been given, in the then state of meteorology. The author has made a large number of unpublished observations on the subject, and he finds a singular uniformity in the isobaric conditions under which this apparent anomaly appears.

In every case which he has examined, the rain with a rising barometer was associated with a peculiar phase of the northerly type of weather. This, as we explained in the last chapter, is a type or spell of weather in which pressure remains constantly high to the north and north-west of Great Britain, while cyclones form over France or Germany.

The character of this phase will be best understood by means of an actual example. In Fig. 86 we give a copy of a barogram in London on April 20, 1877, and underneath the appropriate portion we have marked the time during which rain fell.

Now, at first sight this might seem opposed to all we have said before as to the nature of cyclone-barograms.

Instead of a well-marked fall of the mercury, with rain near the lowest portion, we see a remarkably uniform

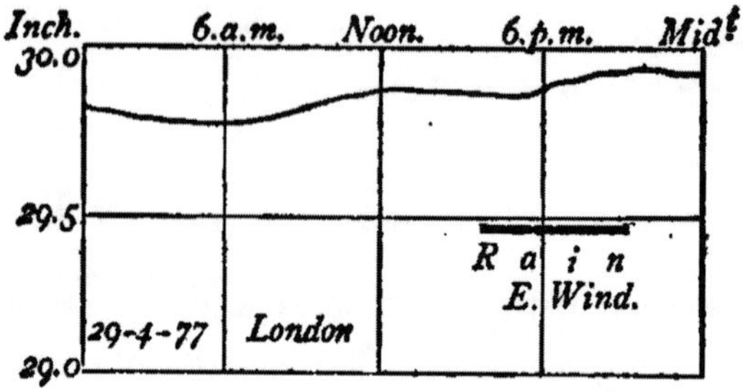

FIG. 86.—Rain with rising barometer and east wind.

trace, in which the diurnal variation of the barometer is very clearly marked. A slight general rise, however,

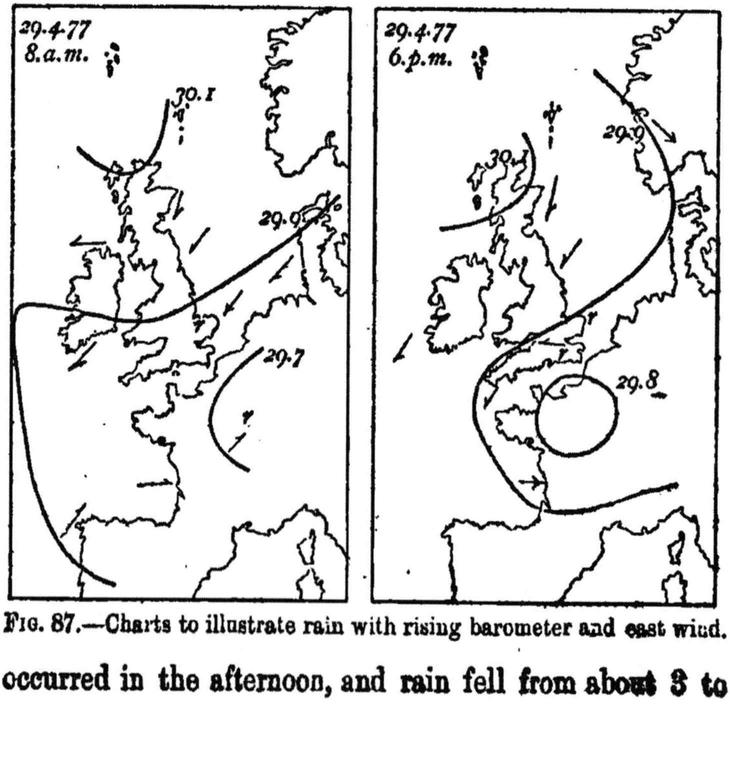

FIG. 87.—Charts to illustrate rain with rising barometer and east wind.

occurred in the afternoon, and rain fell from about 3 to

9 p.m., while the wind remained with little change from north-east.

In Fig. 87 we give charts at 8 a.m. and 6 p.m. on that day, so that the changes of pressure as shown by the isobars may be readily apprehended. In both charts the edge of an anticyclone covers the north of Scotland, but the ill-defined area of low pressure which lay over France and the English Channel in the morning had by the afternoon gathered itself into a well-defined secondary over the north of France. At the same time, partly by a slight general upward surge or increase of pressure—for the lowest isobar in the second chart is 29·8 instead of 29·7—and partly by the advance of the isobar 29·9 nearer to London, the pressure rose in that city, as shown in the barogram, while the rain was due to the formation of the secondary. In this, as in all other similar instances, the advance of pressure from the north-west appears to develop small secondaries, just as a big advancing wave makes small eddies in front of itself.

These secondaries give rain with a rising barometer and east wind.

Rain with Rising Barometer and West Wind.

In the ordinary course of depressions the barometer falls before rain, because the centre of the cyclone contains the rain-area; and if all cyclones moved along a pretty regular path, and did not alter much either in depth or extent, then we should never fail to forecast rain correctly whenever we saw the mercury began to fall, and fine weather soon after the barometer began to rise. But sometimes, before either the centre or trough of a cyclone has reached a station, the depression begins to fill up so rapidly that the barometer actually rises, though in front of a cyclone. Then we get rain with a rising barometer, but the sky retains the appearance due to the front of a cyclone. These are among the cases when synoptic charts enable us to explain what would be hopeless without them, and to see the truth of the statement that weather depends on the position of the observer in a cyclone, and not on the height or motion of his barometer. The next example will illustrate this point very clearly, and also the complications which arise when a secondary forms in rear of a cyclone.

The following sequence of weather was observed by the author at the beginning of September, 1883, about sixteen miles west of the town of Leeds, and 520 feet above the sea-level. His journal, as written at the time, runs thus:—

"*September* 1, 1883.—Early, blue sky, misty, heavy dew, wind south; by noon, sky threatening, halo of 46° diameter, visible from 12 to 2 p.m., then overcast. At 4.45 p.m. light rain began; wind to south-east, almost calm. By 8 p.m. rain heavy; wind up and more to east Barometer fell fast all day. Remark: Very slow coming on.

"*September* 2.—Early, a gale; 8 a.m., uniform nimbus, wind south-east, moderate; the same all day, rain off and on, often rather heavy, but the wind falling light towards night. At 8 a.m. it was seen that the barometer had been falling all night, and was then very low. The mercury continued to fall all day till 6 p.m., when it turned without a squall; though about 6.45 there was a passing shower. After this there was not the look of the rear of a cyclone. Remark: Rain with a south-east wind lasts long; twenty-eight hours.

"*September* 3.—Warm, wet, and stormy; soft; south-south-west to south-west, fresh to strong. About 5.30 a.m. a squall; 6.30, heavy shower, wind round. All day dirty, misty, driving showers, though barometer rising fast. About 5 p.m. rain off, but soft stratus, not cumulus of cyclone-rear; night overcast. Remarks: Weather like front, not rear of a cyclone, and much worse than the two previous days, when the barometer was falling."

By next day, the sky was bright, and covered with cumulus and occasional showers, as is usual in rear of a cyclone, while the wind was round to the north-west. What we have, then, to explain is, first, twenty-eight hours' rain with a falling barometer, and then twenty-three hours' rain, and worse weather after the mercury began to rise; also, not only the increased severity of the weather, but why the sky did not assume its ordinary appearance after the centre of the cyclone had apparently passed.

In Figs. 88 and 89 we give the 6 p.m. chart for September 2, and also that for 8 a.m. the next day, on a large scale; the coast lines are omitted for the sake of clearness, but the position of the letters W, P, D, for Wick, Penzance, and Dover, will sufficiently indicate the scale of the chart. The spot marked L shows the station where the observations were made.

At 6 p.m., September 2nd, the centre of a cyclone of considerable intensity lay near Loughborough, but the isobar of 28·7 ins. had the form of an ellipse, whose longer axis lay from about Barrow to near Cambridge. This gives the direction of the trough of the cyclone,

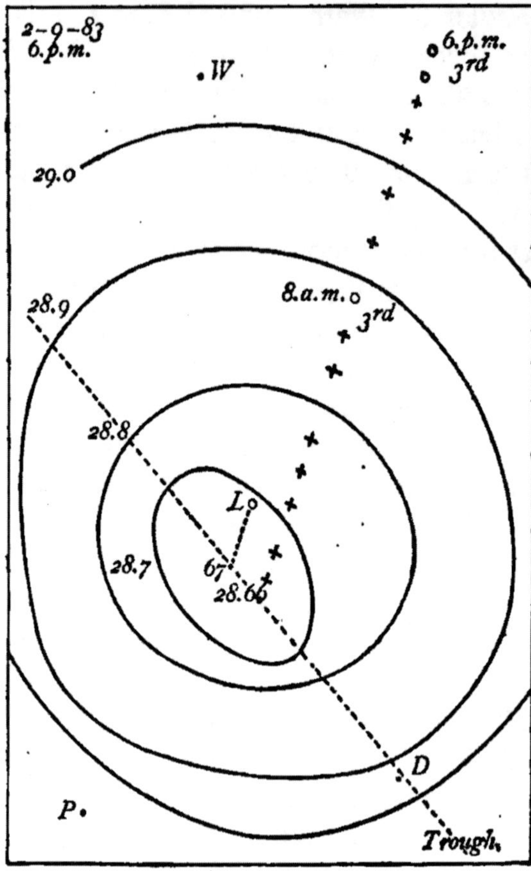

FIG. 88.—Bad weather with rising barometer.

which in this case forms an acute angle of 60° with the direction of the path of the cyclone.

If we assume that the cyclone travelled at a uniform rate till 8 a.m. next morning, it must have been at least three hours before the trough of the cyclone passed near Leeds, that is to say, before the point on the trough marked 0·67 in Fig. 88 reached L. The barogram at L,

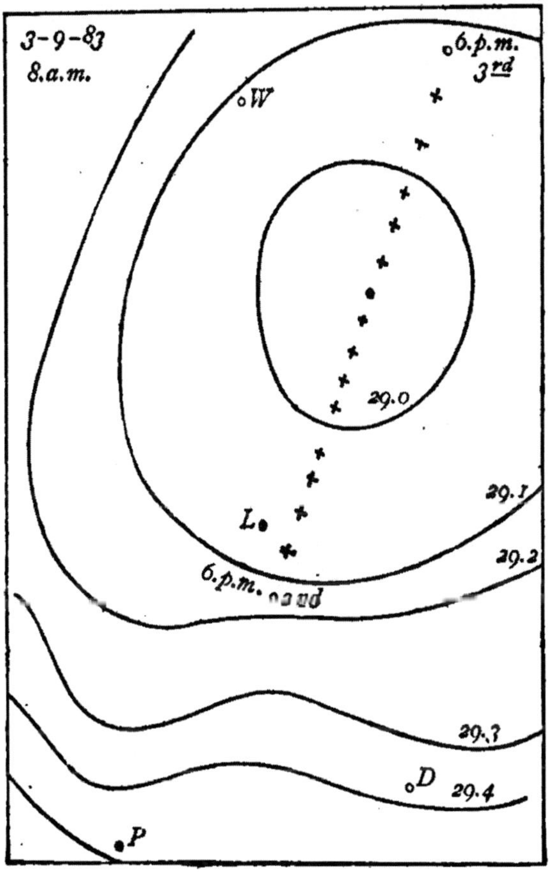

FIG. 89.—Bad weather with rising barometer.

however, began to rise at 6 p.m., and the explanation of this apparent anomaly is as follows:—

At 6 pm. on the 2nd (Fig. 88), the lowest barometer marked 28·66 ins., while by 8 a.m. next morning

(Fig. 89) the lowest pressure was only about 29·0 ins., so that the cyclone had filled up by 0·34 inch during those fourteen hours, or at the rate of 0·024 inch per hour.

It would, therefore, appear that the barometer rose for at least three hours near Leeds, while the centre of the cyclone was still approaching, because the rise of the mercury due to the filling up of the cyclone was greater than the fall of the barometer owing to the approach of the cyclone-trough.

The actual figures in this instance were as follows:—If there had been no filling up, the mercury should have fallen 0·03 inch during the three hours 6 to 9 p.m. This is got at as follows:—The difference of pressure between L, 28·7 ins., and the point on the trough marked 0·67 is 0·03 inch, and the distance between the two points is fifty miles.

This would be traversed in three hours, because the cyclone-centre moved two hundred and thirty miles to the point marked 8 a.m., 3rd (Fig. 88), in the fourteen hours which elapsed between the times for which the charts are constructed.

If there had been no advance, but only a filling up of the cyclone, the mercury would have risen 0·07 inch per hour. Therefore the balance of rise over fall should have been 0·04 inch, and this was the amount actually observed. At 8 a.m., September 3rd (Fig. 89), we see that the centre of the primary cyclone was north-north-east of Leeds, and about two hundred miles distant. The centre was also moving north, so that the motion of the barometer would be upwards from the action of the primary. There is, however, a marked irregularity in the lie of the isobars over the Irish Channel, which points to the existence of a secondary in that neighbourhood. The chart at 6 p.m. the same day showed that the secondary which lay over the Irish Channel at 8 a.m. had become more pronounced, and had then its lowest portion near Liverpool. Consequently Leeds and its neighbourhood were still under the influence of the front of this secondary, though the mercury had risen about 0·2 inch, partly owing to the progress of the primary, and partly also to the cyclone gradually filling up.

By next day the charts showed that the primary had moved still further to the north, and that the secondary was lying over the North Sea, so as to form a sort of V-shaped depression to the south of the primary cyclone.

The explanation of the apparently anomalous weather is then very simple. The first twenty-eight hours of rain with a falling barometer were due to the front of a primary cyclone which was moving very slowly; and so far this represents the usual sequence of weather in such cases. The first three of the twenty-three hours of rain with worse weather after the barometer began to rise were also due to the cyclone-front, though the mercury rose from filling up. The remaining twenty hours of rain and the characteristic sky of the front of a cyclone were due to the formation of a secondary in rear of the primary; so that though the barometer was rising, owing to the passage and filling up of the primary, still Leeds was during the whole of that day exposed to the influence of the front of the secondary with its characteristic dirty weather. The wind was stronger after the glass began to rise, because the gradients were steeper in rear of the primary than they had been in most portions of its front. By the fourth day the secondary had passed away, and then the typical weather of the rear of a cyclone was experienced.

Rain with Steady Barometer.

So far for rain with a rising barometer; now we must consider precipitation with a steady barometer. To Englishmen this is more perplexing than rain with a rising mercury.

In the latter case, we see at once that there is some disturbance going on; but in the former we often have a steady downpour for several hours, with an absolutely steady barometer. Rain of this class is much more common in continental Europe than in Great Britain, except in one very rare case, which will be mentioned hereafter.

The rain is always either non-isobaric, or of that kind which is associated with secondaries and not with primary cyclones. For this reason, the rain is never accompanied by a gale of wind, though there are often angry gusts at the beginning and end of the rainfall.

In Fig. 90 we give a photographic engraving of the author's barographic trace in London, on July 1 and 2, 1877. This, being absolutely untouched by hand, gives the minute irregularities of pressure in a manner which no hand-copied diagram can ever do. The horizontal lines represent differences of half an inch of pressure, the thickest one marking the level of 29·5 inches. The horizontal lines are drawn at six-hour intervals.

At first sight, there might seem to be little sign of any disturbance, for the actual changes of barometric level are insignificant, and the diurnal variation is more obvious than usual in Great Britain. If, however, we look carefully at the trace, we shall find that just before 6 a.m. on July 1 there is a very small dip of the barometer, and that then the trace is almost quite straight till about 4.30 p.m., when there is another small dip; after which the regular diurnal variation is absolutely undisturbed. In London rain commenced at the first dip, and continued without intermission till the second, after which the sky cleared.

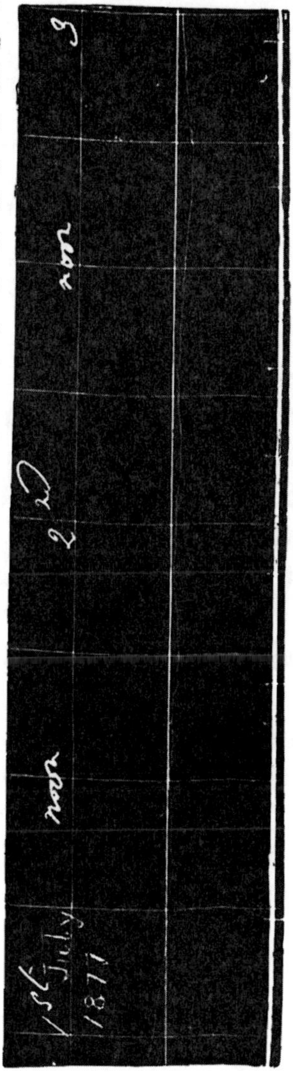

FIG. 90.—Rain with steady barometer.

The charts for that day, which unfortunately the number of illustrations at our disposal does not admit of reproducing, show that this was all caused by the formation and passage of a small secondary over the north of France and the English Channel; and both the rain and the barographic trace are most characteristic of this class of depression. A case of this sort shows, more than any other, the superior value of a continuous trace over an intermittent barograph; for, though the latter permits of the tabulation of hourly values for the determination of diurnal variations, they entirely lose all chance of following the more minute alterations of pressure, which are often accompanied by great changes of weather. The most interesting point about secondaries is the contrast between the intensity of the weather which they induce and the apparently small disturbance of pressure. In primary cyclones the gradients are to a certain extent a good measure of the intensity. In secondaries, on the contrary, the rainfall has no relation whatever to the barometric disturbance. This, of course, makes it very difficult for the forecaster. All he can say when he sees a secondary is—rain; but he can give no estimate of the quantity of precipitation, as he can of the force of the wind in a primary cyclone.

Rarely in Great Britain, frequently in continental Europe, habitually in the tropics, we have purely non-isobaric rains, totally unconnected with any secondary. These are often indicated on the barographic trace by a sudden sharp rise of the type we illustrated in our chapter on Thunderstorms. This is probably a purely local effect of a heavy downpour pressing the air down by its own weight.

The other case of rain—this kind often with a gale of wind—with an apparently steady barometer, only occurs in very unsettled weather. In our chapter on Weather-Types, we gave several examples of nearly stationary cyclones, which increased much in depth, while some of the adjacent anticyclones increased in height. As a necessary consequence, there must be some station where no change of pressure would be observed; but on one side pressure would decrease, while it increased on the other; so by this means very steep gradients might come to lie over the station. The wind would rise to a gale, while the weather would conform to the shape of the isobars, but the mercury would remain stationary; we might, in fact, say that the station was "nodal" as regards the fluctuations of surrounding pressure.

It would be an extreme case when no change of pressure took place, and could only happen at a limited number of places. But under the same conditions there will always be a number of stations where only a moderate fall of the barometer takes place, but a gale out of all proportion to the apparent depression is experienced. This illustrates the important difference between the fall of the barometer due to the passage of a well-defined cyclone, and that due to the rearrangement of the distribution of pressure round the station. As an example, we may turn to Fig. 93 in the next chapter, where we give two charts of North-Western Europe, on February 6, 1883, at 8 a.m. and 6 p.m. respectively. The position of the isobar of 30·4 ins. is practically the same in both maps; but between the morning and evening observations, pressure has fallen 0·4 inch in the west of Ireland, and risen 0·2 inch over Sweden. The shape of the isobars has not altered much, so that gradients have become steep, with little change of wind-direction. Thus many stations, near the nodal isobar, will experience an increase of wind with either a rising, stationary, or slightly falling barometer. For instance, at Aberdeen, marked A, the wind-arrow shows that the wind had risen from a fresh breeze to a moderate gale; while the motion of the isobars does not indicate a fall of more than 0·1 inch in the ten hours which elapsed between the two sets of observations.

Fine Weather with Low or Falling Barometer.

From the above, in which the weather is out of all proportion to the depression of the mercury, we readily pass to the converse case, in which the fall of the barometer is quite disproportioned to the severity of the weather which is afterwards experienced. In the North of Europe, during the winter months, and when the westerly type of weather prevails, the barometer will sometimes fall half an inch or more, and often below 28·5 ins., while no strong winds follow, and the general appearance of the sky is bright, with perhaps a little cumulus cloud. This also is readily explained by reference to our large Atlantic charts. In them we saw that when the Atlantic is covered by a persistent area of low pressure, the depth of the lowest point often suddenly decreases nearly an inch, and that the gradients near the centre are very slight. In some phases of that type of weather, the area of low pressure stretches over Europe, and the *minimum* of this area rises up and down exactly as when the centre lies over the ocean.

If, then, Great Britain, for instance, lay within that area, pressure might decrease a whole inch, and neither storm nor rain be experienced. The great fall of pressure would, of course, develop steep gradients, somewhere to the west of those islands; but as the depression was not caused by the drifting past of a cyclone, neither wind nor rain would follow in England. The centre of these great depressions, which are not true cyclones, is usually associated with cool, bright weather and cumulus cloud, and therefore weather of that description would probably be experienced. From a case of this sort, we learn how to avoid the popular errors that the violence of a gale is always proportional to the fall of the barometer, and that a very low barometer is necessarily associated with very bad weather.

Complications on board Ship.

All the examples which we have now given in this chapter will sufficiently explain the nature of forecasting by means of a single barometer and observations on the appearance of the sky, as also the true nature of the apparent exceptions to the ordinary relationship between weather and the movements of the mercury in a barometer tube.

Our space, unfortunately, does not permit us to describe the still greater difficulties which occur when the observations are taken on board a moving ship; then, of course, we have not only the motion of cyclones, but also that of the ship to take into account, and it is manifest that many of the rules which we have laid down for land-stations would require considerable modification. The same limitation of space also compels us to omit the notice of the theory of handling ships in the small cyclones which occur in tropical countries under the names of hurricanes, typhoons, etc., but the author hopes to make this branch of meteorology the subject of another work.

CHAPTER XV.

FORECASTING BY SYNOPTIC CHARTS.

Statement of the Problem.

By synoptic forecasting we mean that branch of weather-prevision which is carried on by means of synoptic charts. The forecaster in a central bureau is in telegraphic communication with observers for many hundred miles round. From their reports he constructs synoptic charts at such intervals as seem necessary. To the indications which he derives from the appearance of these maps, he adds all his own accumulated experience of the nature of the meteorology, and the motion of depressions in his own country; and also such knowledge of the recurrent periods of different kinds of weather as he may be acquainted with. From all that he forms his own judgment as to what changes are likely to take place, and issues his forecast accordingly.

From the nature of things there can never be many forecasters. The rapid nature of meteorological changes makes the employment of the electric telegraph absolutely necessary, and the great expense which is thereby incurred, compared with the uncommercial nature of the results, practically relegates forecasting to the functions of a Government office.

From the preceding chapters we now know what weather is. Instead of dealing with abstractions called wind, rain, cloud, heat, etc., we have gradually been led up to the idea that all meteorological phenomena are the products of the motion and circulation of a moist atmosphere. Now we know that when we talk about forecasting weather, we mean that we are going to say how or where certain aerial eddies will move, or when new ones are likely to form; also whether any cyclone will be violent or gentle.

AIDS TO FORECASTING.

In this chapter we propose to make some additional remarks on the whole aspect of the subject. We shall enumerate several aids to forecasting which can be obtained from various sources, and point out both the present difficulties and the future possibilities of weather-prevision. Finally, we shall give some examples of successful and unsuccessful forecasts in different countries, and an account of the various percentages of success which the different offices have achieved. In an international work we shall better illustrate the general principles of the subject by exemplifying forecasts in different countries than by trying to give any one in detail. A tolerably full account of the nature of forecasting, and of the details of the methods and machinery for issuing storm-warnings in Great Britain, will be found in the author's work, "Principles of Forecasting by Means of Weather Charts," issued by the authority of the Council of the Meteorological Office.

Unequal Barometric Changes.

We have already fully explained the use of the recognition of weather-types in every country, during which sequence the motion of depressions follow either a certain general direction or maintain a certain general position; but in variable climates we often find tracts of weather which can be assigned to no particular type. The forecaster is then at a great disadvantage, for he has little to guide him as to the future.

The very idea of weather-type involves the knowledge that the sequence of changes will follow in a certain groove, so that when no type is obvious, there is little basis on which to frame a forecast. In most cases the forecaster has to rely on the difference of barometric rate in various districts. If he sees that the barometer is falling much more rapidly in one district than in any other—even if no definite depression is formed—he knows that steeper gradients must thereby be formed, so that the wind must increase, and whatever weather is due to the existing shape of isobars will get worse.

Conversely, if he finds pressure increasing in a district of low barometer, he knows that gradients will decrease, and that both wind and weather will moderate. The details vary indefinitely, and no rule can be laid down even for a single country; everything must be left to the judgment and experience of the forecaster.

CYCLONE-PATHS.

The paths of cyclones, and the nature of the influences which deflect or otherwise alter them, are so important that we propose to devote some paragraphs to their consideration, of course with a special reference to the bearing which they have on forecasting.

When the paths of the rare but violent cyclones of the tropics, which are known as hurricanes, typhoons, or cyclones, are plotted on a chart, we find that, though there is a general similarity in their tracks, there is still so much difference that we cannot attempt to lay down any absolute law of their motion.

For instance, the West India hurricanes usually begin with a westward course, and then gradually bend round till they end by moving towards the east or northeast. But in some instances they continue in a westerly direction, and traverse the southern portion of the American Union, instead of curving round across the Atlantic.

For this reason, if a ship was handled on the supposition that the hurricane would always go the same course, she would be exposed to very great danger.

In the temperate zone, where cyclone-paths are still more irregular, any attempt to lay down any hard and fast rule for the tracks of depressions could only lead to disastrous failure of any forecasts which were based on that system; but though the numerous causes which have been found to modify the paths of cyclones cannot be allowed for in estimating the probable future path of any actual depression, still many points, which have been noted, are so interesting that we shall mention some of them more in detail.

Tendency to follow certain Tracks.

During the persistence of any type, two or three successive cyclones have a remarkable tendency to follow the same course. This, of course, is the natural product of the fact that the path of a cyclone is determined by the type of pressure in which it is formed. Sometimes this path is entirely dictated by surrounding pressure; but at other times local configuration of the land exercises a most powerful directive influence.

For instance, in Great Britain, during the westerly type, when the depressions are so far south as to cross that island, the centres have a decided tendency to traverse either the line of the Caledonian Canal in Scotland, or the low-lying ground which separates the valleys of the Forth and Clyde. Both of these courses coincide with what we may call lines of least resistance, for these are the two easiest lines by which it is possible to cross the mountainous districts of Scotland. Another well-marked tendency of cyclone-centres is to hug the sea-shore, rather than to strike inland. When a cyclone comes up the English Channel, it often skirts the south coast of England, and then moves more northward along the east coast, rather than pass directly to the north-east across the land. In like manner, large cyclones which come in from the Atlantic, when they meet the coast of Norway, often hug the coast for several days, instead of going straight to the north-east. In the United States the great majority of cyclones traverse the line of the great lakes, and then either follow the valley of the St. Lawrence or strike across the New England States into the Atlantic.

Great chains of mountains also influence very powerfully the paths of cyclones.

In Europe, the chain of the Alps almost forms a natural boundary between the weather of the Mediterranean and that of the northern portion of the continent. As a rule, that great inland sea has a totally different atmospheric circulation from that which affects the rest of Europe.

This will be very obvious if we turn again to the large charts which we gave in our chapter on Weather-Types.

Sometimes we can trace a cyclone in the Mediterranean trying to cross the Alps, and being broken up in the attempt. We can readily understand that if a mountain chain, 12,000 feet high, sliced off the lower half of such a shallow and complex vortex as a cyclone, the whole system might very easily be destroyed. Exceptional cases, however, do occur in which large cyclones cross the great barrier of the Alps.

In India, too, the still loftier chain of the Himalayas imposes an even greater influence on the meteorology of that country, as a glance at the charts which we have already given of the monsoon districts will abundantly show.

Storms crossing the Atlantic.

But the cyclones whose motions have created by far the greatest interest in Europe are those which sometimes come across the Atlantic. The public have been fascinated by the idea that a storm could be telegraphed from New York, and its arrival on the coasts of Europe foretold three or four days in advance. If cyclones only moved with tolerably uniform velocities and in tolerable uniform paths, and the intensity remained constant, then, indeed, it would often be possible to obtain timely warning from the United States or Canada. Although the diagrams which we have already given of Atlantic weather would sufficiently show the real character of Atlantic cyclones, still the nature of the paths of these depressions will be more clearly understood if we give the tracks of all the depressions which appeared in the Atlantic during a single month. This will do as a sample of any other month or season. In Fig. 91 we therefore give a chart of all the cyclones which could be traced for more than two days in the United States, the Atlantic, and Europe during the month of July, 1879. During that month there were seven well-defined cyclone-tracks within the above-mentioned area.

These paths are plotted on our chart, and the position of the centre of each cyclone on every day is clearly marked.

Now, the first glance will at once satisfy us as to the broad idea that cyclones usually move in a certain general direction.

The whole of the paths lie along a comparatively narrow belt of the ocean; but when we come to look into the details, we shall find that the smaller variations of motion effectually preclude the use of this knowledge in forecasting.

Of the seven cyclones, four—Nos. I., II., V., and VII. —were formed in mid-Atlantic, and then pursued a more or less irregular course towards Europe. Observe how the curious loop to the northwards, which the path of No. I. makes at the beginning of the month, is almost exactly reproduced at the end of that time by cyclone

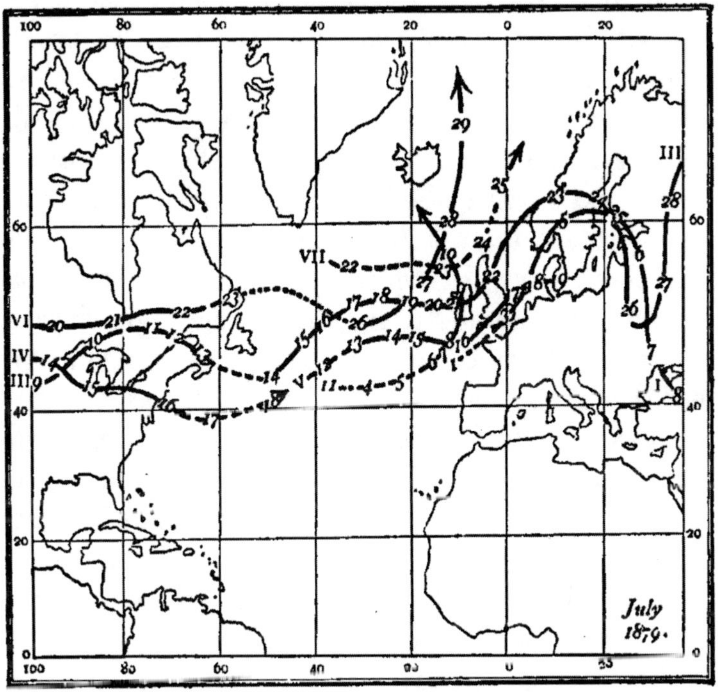

FIG. 91.—Cyclones crossing the Atlantic.

No. III. Cyclones IV. and VI. were formed over the United States; both passed into the Atlantic, but neither reached the coasts of Europe.

Cyclone No. III. also had its origin in the American Union, though, unlike the two others, it not only survived its journey across the Atlantic, but, after traversing Europe, passed into Siberia. Our chart follows its history for the ten days from July 9 to 28; but let us try to see how we should fare if we attempted to issue forecasts on the supposition that the depression would move either in a uniform direction or with a uniform velocity. From the 9th to the 11th the cyclone moved towards the north-east with a considerable velocity; the next two days it turned to the south-east with diminished speed, and left the shores of the United States with a south-easterly trajectory. The day of leaving the velocity increased; but by next morning the direction changed again to the north-east, and the velocity gradually diminished for the next seven days, by which time the depression had reached the coast of Ireland, after being eight days in transit from Nova Scotia. A crack steamer would have done the distance in five days. From that day, the 21st, the speed increased again, and the cyclone turned still more towards the north. Then, with gradually decreasing velocity, the path bent round to the south, and afterwards turned once more to the northwards, with increased speed, till the 28th of July, when we lose sight of the depression in the frozen marshes of Siberia.

This example will abundantly prove that we can form no estimate of the future path or velocity of a cyclone-centre by any observations on its earlier motion. In this case the direction and velocity of the depression when it left the American shore gave no clue either to its path across the ocean, or its meanderings after reaching the continent of Europe.

There is another point which we must remember in the discussion of this question—we track cyclones, but not necessarily storms. The size and intensity of this cyclone varied every day of its life. Some days the intensity was so great that the wind rose to the force of a gale in places; other days the gradients were never developed of sufficient steepness to give rise to more than a breeze. No general rule can be laid down that will apply to the life-history of a cyclone; we must watch from day to day for symptoms of increasing or decreasing intensity.

From all this we can also estimate the value of the idea that a swift Atlantic mail-steamer could arrive before a storm, and so give notice of approaching danger.

The cyclone which we have just traced travelled rather slower than usual; we often find depressions cross the Atlantic in four days. However, in this case, the cyclone came across at just about the speed of the fastest steamers. The first two days the cyclone would have been passing the vessel; on all the other six days, the steamer would have been catching up the cyclone. The ocean route from the mouth of the St. Lawrence to Cork is almost exactly along the track of this cyclone. A steamer would, therefore, have experienced little wind, but a uniformly low barometer during her voyage. Any report which she alone could give would be useless to a forecaster in London or Paris; but if several boats were arriving, and they all telegraphed up their observations at 8 a.m. on the three or four preceding days, then the combination of their results would certainly enable the forecaster to deduce some useful indications.

In all British forecasting a certain amount of uncertainty must always remain as to the future path of a cyclone, even when we see a well-defined depression lying off the coasts of Ireland; how much greater must the uncertainty be when we attempt to forecast the path of a cyclone four days ahead, and from a distance of three thousand miles? If the forecaster cannot hit England straight when he aims from Ireland, will he be likely to hit her at all if he shoots from New York? The number of cyclones which actually cross the Atlantic from shore to shore appears to vary from about eight to twenty in any year. In many cases it is difficult to say whether it is the same cyclone which we trace, from the peculiar manner in which two depressions may fuse into a single new one. On the whole, then, we see that the crude notion of forecasting European storms from the United States contains some elements of truth, but that still, from the nature of cyclone-motion, the idea can never be used in practical forecasting.

Path as indicated by the Strongest Wind and Highest Adjacent Pressure.

A good deal of work has been done, both in England and Germany, on the question of how far the path of a cyclone can be determined by the general direction or force of the surrounding wind, and the investigators have found that generally the propagation of the cyclone is in the same direction as the strongest surface-wind in the neighbourhood. There are, of course, a good many exceptions; and it is impossible in our present state of knowledge to say whether the strongest wind indicates the general direction of the generating current in which the cyclone is only an eddy, or whether the strongest wind is the product of the combination of surface rotation and propagation, being nearly in the same direction at one particular point.

All our charts have shown that a cyclone usually tries to keep an area of high pressure on its right-hand side; and this, too, has a good deal to do with the strongest wind being found at right angles to the centre, and therefore nearly in the same direction as the motion of the whole depression.

Influence of Surrounding Temperature.

We now come to the far more difficult but important question as to the influence of surrounding temperature on the propagation of cyclones, and as to whether the development of heat on the right-hand side of a cyclone is the cause or product of cyclone-motion.

Putting all theoretical considerations aside, the facts of the case, as far as Europe is concerned, are as follows:—A cyclone nearly always has the highest temperature on the right-hand side of the path; and for the same distribution of pressure, there is a considerable difference in the path of depressions at different seasons of the year, when the general slope of heat from the equator to the pole is not the same.

Dr. J. V. Bebber has discovered the following relations special for Germany and Central Europe:—"If the distribution of air-pressure and temperature in the neighbourhood of a depression are directed to the same sense, then the propagation of the depression is nearly perpendicular to the pressure and temperature-gradient. If the air-pressure and temperature in the neighbourhood of a depression are distributed in an opposite sense, and if the differences are nearly equal, so is the motion of the depression checked, or even arrested (stationary depression), whereby the depression takes a long, more or less distorted form, of which the longer axis lies perpendicular to both the air-pressure and temperature-gradient. If, with the same distribution as before, either the air-pressure or temperature-gradient overweighs on one side of the depression, so will the direction of the path be determined by the predominating element. If air-pressure and temperature are not, indeed, opposite, but also not distributed in the same sense round the depression, so will the depression strike out a resultant direction." He also thinks that pressure is the more important determination of cyclone-motion in winter, and temperature the predominant influence in summer.

The conception of temperature and pressure gradients being distributed in the same or opposite senses, appears to be as follows:—If the highest pressure and highest temperature are either both to the north, or both to the south of a cyclone, they are said to be in the same sense, and the depression will move at right angles to both. But suppose pressure was highest to north, and temperature to south; then these two elements would be distributed in the opposite sense, and the cyclone would probably be arrested in its usual eastward course.

These observations are more suitable to Germany than to Great Britain, as some of the expressions are hardly applicable in the latter country, and in England local variation is so great, and the area of observation so small, that the distribution of surrounding temperature can scarcely be used in practical forecasting. But in all continental Europe we have one practical rule—that if pressure is high to the north or north-north-east of a cyclone, and temperature also higher on that side than to the south, then the propagation of the depression will probably be towards some point of west, instead of towards the east as usual. For instance, suppose we found some morning a cyclone over Central Europe, with an anticyclone over the North Sea, the natural presumption would be that the depression would move—always very slowly in this type—towards Russia; but if, as in Figs. 95 and 96, we found the highest temperatures in the Baltic, and not in Austria, and especially if the temperature seems to rise to the north or north-west of the centre, then we might forecast that the depression would move, as in this instance, westwards towards Great Britain.

The question how far the cyclone affects temperature, and how far the latter directs the former, will be best explained as follows:—Let us call the general slope of temperature from land to sea, which varies according to the time of year, the "seasonal gradient of heat," and the patch of heat on the right of a cyclone "cyclone heat;" then we may say that, while the seasonal gradient has a directive influence on the path of the depression, the cyclone heat is the product of the moving whirl itself. The conclusive proof that the heat-patch on the right front of a depression belongs to the cyclone directly, and not indirectly, through the disturbance of radiation, is found in that peculiar quality that no thermometer can appreciate, but which is readily recognized by our more delicate sensations. In a typical east-going cyclone the neuralgic, pain-producing heat comes with the south-east wind on the right front of the depression; but when a cyclone goes west, the then right front has a north-west wind and the same distressing quality of heat.

Forecasting depends on no Theory.

We can now readily understand from all the foregoing remarks that forecasting depends neither on any theory nor on any calculation. The whole science, from beginning to end, rests solely on observation.

The shapes of isobars, and the relation of wind and weather to them, are matters of experience only. We find that certain kinds of weather are associated with different portions of each fundamental form of isobars and we classify accordingly. We give each shape of isobars a conventional name, but that does not bind us to any theory of atmospheric circulation. In like manner, we see that no averages or mean values are of any avail in forecasting weather. Cyclones may usually take a certain path, but they need not do so; the greater portion of the rainfall of any country may come with a south-west wind, but that does not prevent many fine days with the wind from that quarter. On an average, in England, three days out of four may be cloudy, and the forecaster who always announced a cloudy day would have seventy-five per cent. of success. Still, in an anticyclonic period his calculations would totally fail; he could never say what kind of cloud would appear, and such a system would have no claim to be called forecasting in the modern sense of the word. It is impossible to suppose that we have yet nearly reached the highest perfection of which forecasting is capable, but still we know enough of the nature of the subject to say with certainty that calculation will never enter much into the science of weather-prevision. Natural aptitude and the experience of many years' study are the qualifications of a successful forecaster. "In fact, meteorology is not an exact, but an observational science, like geology or medicine; and just as, however accurately the symptoms or treatment of any malady may be described, the skill to recognize and the judgment to treat must rest on the ability of the physician, so in meteorology, however carefully the relation of weather to isobars may be defined and the nature of their changes described, the judgment which experience alone can give, to enable a warning to be issued, must ever depend on the professional skill of the forecaster."

Detail possible.

It may not be out of place to introduce here a few remarks as to the amount of detail which it appears possible to give to daily forecasts. Under various headings, we have already discussed the influence of local obstacles in modifying the appearance or intensity of any kind of weather, and also the powerful diurnal variations of every element in all parts of the world. When to these we add the tendency of cyclones to form secondaries, so small as not to show in an ordinary synoptic chart, then we may easily understand that it is the general character only of weather which a forecaster can ever safely predict. The general character is the quality of weather which we have taken such pains to show is constant in each portion of every shape of isobars, and that never changes under any local or diurnal variation.

If we live in any place which commands a view over any large tract of country, and we think how often we see both cloud and rain which only affect a very small portion of our horizon, we can readily understand that, even if it were possible to issue minute forecasts, every few square miles of country would require a separate warning.

How far in Advance can Forecasts be issued?

We may also consider how far in advance forecasts can safely be issued. The numerous charts which we have already given will show the reader the amount of change which twelve or twenty-four hours may develop in the distribution of pressure. Sometimes we have been able to trace the changes in either of these intervals quite easily; at other times it has been difficult to say how the first set of isobars has grown into the second. In the United States the observations are taken three times a day, and this appears to be sufficiently frequent for all practical purposes. In most European countries, reports are not sent up more than twice a day; but with this interval, cyclones sometimes form so suddenly that they are not forecast in time to give any warning. We shall give an example of such a case further on in this chapter.

Thus, from eight to twelve hours seems to be the furthest time for which forecasts can be issued in advance, and even then many local details cannot be given. Some meteorologists are of opinion that a good deal of forecasting will be done in the future, with the assistance of a complete knowledge of recurrent periods of heat, cold, rain, or storm; and we lean strongly to that view, if these periods are used in the manner so fully explained in our chapter on Seasonal and Cyclical Periodicities.

Time of Preparation.

A few particulars of the time necessary for collecting and examining the materials for synoptic charts will perhaps enable the public better to understand the practical conditions of the problem of weather-forecasting and storm-warnings.

In Great Britain, the morning observations are taken at 8 a.m. Even with all the rapid organization of the British Post-office, the majority of the reports do not arrive till between 9 a.m. and 10 a.m. As fast as they arrive, the information is entered on a chart, and a synoptic chart is constructed. If necessary, telegraphic intelligence of storms is immediately sent to the coasts, and in every case information as to the state of the weather, and a forecast for twenty-four hours ahead, is sent to the press.

In practice, storm-warnings can rarely be despatched

before 11 a.m.; that is to say, three hours after the observations have been taken. If we allow at least another hour before the public can have access to the information, we see at once that the day is so far gone that the forecast can have little practical importance for the majority.

The greatest value is when a storm has just begun to show over Valentia at 8 a.m.; then the English coasts can be warned in time. Still, in the three or four hours which must elapse before the storm can be warned, the cyclone will have advanced, perhaps, as much as a hundred and twenty miles, so that, before a telegram can reach the western shores of England, the gale will either have commenced, or the appearance of the sky will have given unmistakable warning.

The whole theory of storm-warnings by means of the electric telegraph is based on the supposition that the message travels faster along the wire than the storm along the earth's surface. But, as the practical organization of collection and distribution of intelligence takes at least three hours, the storm must either move slowly or over a considerable intervening district before any set of stations can be successfully warned. The forecasts which are issued from reports taken at 6 p.m. are of more use.

The organization of the press enables the public to obtain the office forecast much more quickly than by any other means. The British reports are taken at 6 p.m., while the United States Signal Office obtain their latest about 11 p.m. These five hours are an unquestionable gain. In Great Britain there are, however, difficulties in the way of transmission of intelligence during the night.

Except in the large towns, the majority of telegraphic stations are closed till 8 a.m., so that while the evening forecasts do not reach them till past eight o'clock in the morning, the information for that morning arrives about three hours later. Thus we see the practical difficulties in the way of forecasting. There is no doubt that in time some of them will be successfully overcome.

When most successful.

A few remarks on the circumstances under which the most successful forecasts can be issued will also much help a general apprehension of the subject. We will confine our observations to Great Britain only. It is very obvious that the more striking the weather-changes, the more have we something definite to forecast. When we have a well-formed cyclone, which traverses a well-defined path, we have strongly marked sequences of wind and weather, and any error in the forecast will only arise from some slight difference between the expected and the actual track. But when we have what we have seen is the more usual state of things in Great Britain—ill-defined depressions which move irregularly, and one or more of which fuse into a fresh cyclone with a new centre—then we have no definite sequence of weather to deal with, but a change which is produced by the weather at each station gradually conforming to the varying shapes of isobars. The best that can be done then is to forecast generally broken weather, and more or less rain generally; but no attempt can be made to foretell any definite series of wind-shifts, as in a true cyclone.

Experience has shown that in Great Britain no serious gale has ever been experienced, unless there is more than half an inch of difference of barometric pressure between some two stations. Synoptic charts will always detect even much smaller differences; so that, though some uncertainty will always remain as to the direction of the wind, the force will generally be at least approximately forecast correctly, except in the case of a very sudden and unexpected fall of the barometer.

Very different, however, is the case of rain. Secondaries and non-isobaric rains are the forecaster's bugbear; they form so quickly, show so little on a synoptic chart, and move so irregularly, that rain in general terms is all that the forecaster can usually say. In summer, when he sees the characteristic loops in the isobars which constitute secondaries, he can safely predict thunder and rain; but he cannot attempt to localize either of these phenomena.

Sometimes, too, secondaries are so small that they do not show at all on a synoptic chart, which is constructed on reports received from stationsoften a hundred and fifty or two hundred miles apart. The whole loop of a secondary need not be nearly so large; and then a depression of that class might lie between two stations, and yet be indicated at neither. The weather, however, would be profoundly modified, and the forecasts would probably be erroneous.

There is also always the important difference between wind and rain, that the former is always in the main determined by the steepness of the gradients, while the amount of precipitation bears no relation to any known meteorological element.

In many shapes of isobars we know that there will be rainfall, but whether much or little, we cannot tell at present.

From these considerations we need not be surprised to find that in all offices, except in Japan, wind is better forecast than rain.

SOURCES OF FAILURE.

From the conditions of successful forecasts, we can readily turn to those of unsuccessful predictions. Besides the uncertainty of rainfall due to the action of secondaries, there are four principal sources of failure: the sudden formation of an intense cyclone; the sudden dying out of an existing cyclone; the motion of a cyclone in an unexpected path; and, lastly, an error in the judgment of the forecaster.

In the first case, of the sudden formation of a new cyclone, the whole forecast is necessarily totally upset, and the weather which is experienced is worse than had been anticipated.

The converse occurs when an intense cyclone suddenly dies out. Then the weather is much better than was expected; but neither in this case nor in the preceding one can settled weather be expected.

When a cyclone takes an unusual path, the general character of the weather will remain bad, but the direction of the wind and the details in different districts will be wrongly forecast. We have already given instances of cyclones which move in no well-defined path, and more complicated cases often occur. Sometimes the path will describe a complete circle of no very great diameter; but the commonest case in Western Europe is when the path of a cyclone takes the form of the letter V. For instance, a cyclone comes in from the Atlantic from about due west, and after it has gone as far as England, it moves back again in a north-westerly direction, as it has not been able to pass the area of high pressure which would then be lying over Northern and Central Europe. In another common case, the cyclone comes down from the north-west on to England, and then passes off in a north-easterly direction towards Norway. In all such cases the forecaster is at a great disadvantage.

Lastly, the judgment of the forecaster will sometimes err. We have shown that no absolute law of cyclone-motion can be laid down, and that, in fact, the tracking of well-defined depressions forms but a small portion of the forecaster's business. On the larger number of days he has to estimate how, or where, cyclones will form in an ill-defined area of low pressure, or how far an area of low pressure will encroach on another region of high barometer. In this, he must rely on his own opinion and experience alone; that must be fallible sometimes, but better results are obtained by trusting to personal skill than by attempting to use any mechanical rules or maxims.

Men differ in their aptitude to forecast weather in the same way as physicians differ as to the accuracy of their diagnosis; but just as the best results are obtained by selecting the doctor whom experience has shown to be the most successful practitioner, so the best forecasts are got by selecting the meteorologist who has been the most successful in that branch of the subject. In the United States Signal Office at the present time, four men take the duty of forecasting in rotation. They have so far all been ground in the same mill, by passing through a two-years' course of the same hard training; and it is found in practice that the difference between the best and worst is two per cent. in the number of successful forecasts. For instance, if the best man gets ninety per cent., the worst will attain to eighty-eight per cent. of success.

Some Countries easier than Others.

From all that we have now explained, it will be very evident that forecasting is much easier in some countries than others. In the tropics, the great seasonal changes come on regularly, and the smaller changes from day to day are insignificant. In the two or three days of any year on which a regular cyclone may form, the premonitory symptoms are so obvious that there is no difficulty in framing a forecast.

In temperate regions, those countries will be the best situated which lie to the east of a well-observed land area, because most disturbances in the temperate zone move from the west.

Thus Germany and Norway are much more favourably located for weather-prevision than either England or France.

In the year 1869 twenty-three storms were felt in

Hamburg, and of these twenty-two had previously passed over some part of Great Britain. In the seven years, 1867–1874, 301 warning messages were issued from London to Hamburg; seventy-two per cent. of these warnings were followed by gales, while in only three cases did the storm outrun the message. Then in the United States, the majority of cyclones commence in the Rocky Mountains; so that with the admirable organization of the Signal Office, timely warning of serious gales can usually be sent to the Eastern States of the Union.

Great Britain is situated in a region of peculiar difficulty. Not only does her insular position preclude any early knowledge of the advent of cyclones, but, from the nature of weather-types, she is more exposed to unsettled weather than any other part of Europe.

We have seen in our chapter on Weather-Types, that the positions of the great areas of high and low pressure are to a certain extent determined by the areas of land and water.

When the persistent anticyclone of the southerly type lies over Scandinavia, the Atlantic is covered by low pressure and bad weather; when the great anticyclone covers the Atlantic in the northerly type, then pressure is lowest and weather worst in Scandinavia; so that, in almost every case, Great Britain is on the boundary between a cyclonic or anticyclonic system, and is therefore exposed to changeable weather. Just as an outlying rock is exposed to the wash of every sea, so England is exposed to the disturbing influences of every type of European or Atlantic bad weather.

EXAMPLES OF ACTUAL FORECASTS.

BRITISH.

After these explanations we will now give some examples of actual forecasts in different countries, commencing with Great Britain. The latter are taken, with some important additions, from the author's work on the principles of forecasting before mentioned. We have selected our first example to illustrate a completely successful forecast which depended on the estimate of the forecaster as to the progress of an ill-defined area of low pressure towards the east. This is one of the commonest cases which occur in Great Britain. The chief points which the forecaster had to consider were the direction in which the depression would move, and especially how far east it would pass without being arrested in its progress. Also, whether the gradients would become sufficiently steep to give rise to serious gales.

But to understand properly the details of the warnings, we must first explain the districts into which the United Kingdom is divided for the localization of weather-forecasts.

In Fig. 92 we give a map of the eleven districts in the British Islands which are separately warned; and by means of this map the subsequent details will be easily followed.

A glance at the relative size of any one of these districts and the area covered by even a small cyclone, will show at once how much a small change in the cyclone may mar the most carefully drawn deductions of the forecaster; the smallest loop in the isobars which we saw so often in our large charts of weather-types would entirely alter the details, though not the general character of the weather which would be experienced. The action of such a secondary might reduce the force of the wind so much that some district would receive a warning which was not

FIG. 92.—British forecasting districts.

justified by the event, or develop rain where fine weather had been anticipated and forecast.

In the left-hand portion of Fig. 93 we give the chart from which forecasts had to be issued at 8 a.m., February 6, 1883. We see in it at once the commonest features of the southerly type of weather with the pressure high over Scandinavia and low over the west of Ireland, while the isobars run nearly due north and south. Southerly gales have already commenced in the west and north, while fine weather prevails over the south and east coasts of Great Britain.

It was also known, by comparison with the previous charts, that while the barometer was rising over Norway, it was falling, but only slowly, over the western coasts of Ireland. Now, from all that we have already explained

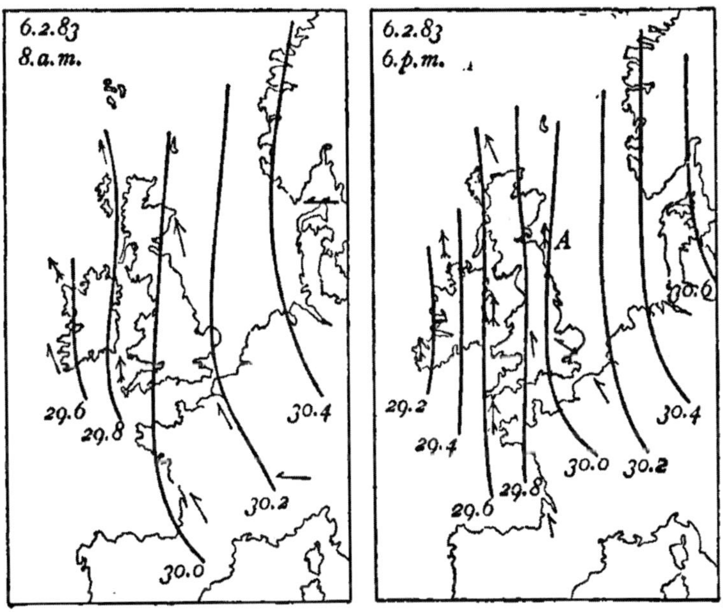

Fig. 93.—Successful forecast (British).

as to the nature of this type, it is evident that there is no fear of the depression crossing England so as to bring any great change of wind, but that the gradients will get steeper for southerly winds with bad weather, and that probably the south and east coasts will not be affected. Then, as to storm-warnings, all the north and west (Districts 0, 1, 6, and 9) were already warned, but as the south of Ireland (District 10) will be affected by the increasing gradients, warnings are now necessary for it also. Hence the following forecasts were issued to the different districts:—

FORECASTS FOR THE TWENTY-FOUR HOURS ENDING AT NOON ON FEBRUARY 7, 1883.

Districts.	Forecasts.
0. Scotland, N. ...	Southerly strong winds and gales; cloudy generally, with some rain.
1. Scotland, E. ...	Do. Do.
2. England, N.E. ...	South-easterly winds, moderate inland, strong on coast; fair generally.
3. England, E. ...	Do. Do.
4. Midland Counties ...	Same as No. 5.
5. England, S., and the Channel	South-easterly and southerly winds, moderate or fresh; fair generally.
6. Scotland, W. ...	South-easterly and southerly strong winds, perhaps a gale; fair to cloudy, and unsettled.
7. England, N.W. ...	Do. Do.
8. England, S.W. ...	South-easterly and southerly winds, increasing; cloudy.
9. Ireland, N. ...	South-easterly and southerly winds, increasing to a gale; cloudy, unsettled; some rain.
10. Ireland, S.	Do. Do.
Warnings ...	The south cone is still up in Districts 0, 6, 9, and parts of 1 and 7, and has been rehoisted this morning in District 10.

By looking at the right-hand portion of the chart (Fig. 93) for 6 p.m. on the same day, we find that the above anticipations have been completely verified. Wind and rain have increased in the west and north, but in South-east England the weather remains fine. In his journal near Dover, in District 5, on that day, the author finds the following entry:—"*February* 6. 1883.—Cold, dry, very fine and bright; wind south-east, fresh." Hence the forecasts were a complete success. The weather was cool near Dover because that town was under the influence of the European anticyclone; but in all the western districts temperature was very high for the season. In the

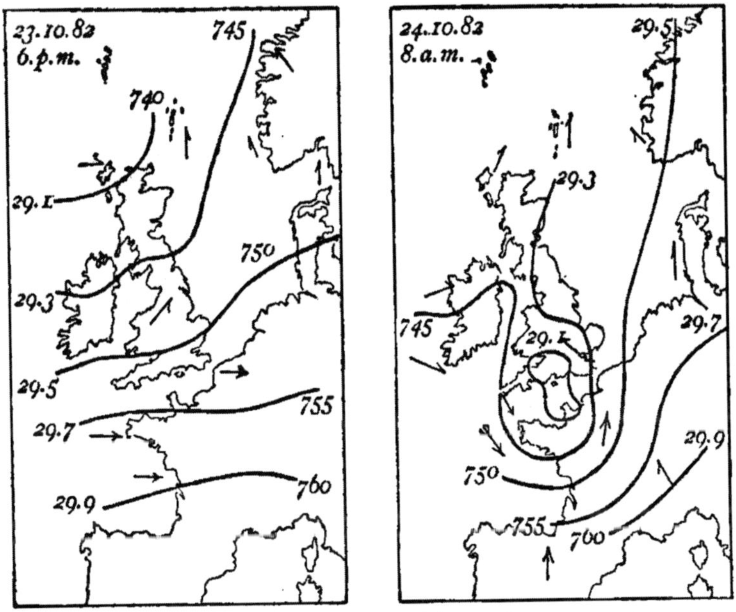

FIG. 94.—Failure of forecasts.

selection of this example we had, however, an additional object, viz. to illustrate what we have laid down relative to the use of periodicities in forecasting.

We have already mentioned that the period February 7–10 is one of recurrent cold weather, whence, if the forecaster had trusted blindly to periodicities, he would have made a complete failure. On the other hand, had he discovered on this day the commencement of either the northerly or easterly types, the knowledge of the periodicity would have been of great use to him.

Our next illustration will be that of a kind which, fortunately, rarely occurs, viz. the sudden formation of a cyclone in an unexpected position, which entirely upsets all forecasting. In the left-hand portion of Fig. 94 we give a chart for 6 p.m., October 23, 1882. There we see the most familiar features of the westerly type of weather, and though the barometer was falling over the Bay of Biscay, and rising over Scotland, there was no reason to expect that the ordinary sequence of that kind of weather would be disturbed—that is to say, that west and south-west winds, with rather showery weather, would prevail. Accordingly the following forecasts were issued:—

FORECASTS OF WEATHER FOR OCTOBER 24, 1882, ISSUED AT 8.30 P.M. THE PREVIOUS DAY.

Districts.	Forecasts.
0. Scotland, N. ...	South-westerly breezes, fresh or moderate; showery.
1. Scotland, E. ...	South-westerly breezes; moderate; some showers, with bright intervals.
2. England, N.E. ...	Do. Do.
3. England, E. ...	Same as No. 5.
4. Midland Counties ...	Same as No. 1.
5. England, S. ...	Westerly and south-westerly breezes, light to fresh; fine and cold at first, some local showers later.
6. Scotland, W. ...	Same as No. 0.
7. England, N.W. ...	Same as No. 0.
8. England, S.W. ...	South-westerly winds, fresh to strong; showery.
9. Ireland, N. ...	Wind returning to south-west, and freshening; weather showery.
10. Ireland, S.	Do. Do.
Warnings ...	None issued.

When we come to look, however, at the right-hand chart in the figure for 8 a.m. the following morning, we find that a small well-defined cyclone had formed during the night over the English Channel, which moved during the day towards north-north-east, and thereby produced continuous rain with complete shifts of the wind through 180° in many parts of the country, so that the forecasts issued were a complete failure.

Present Results.

It will now be interesting to give some idea of the amount of success which at present attends both every-day weather-forecasts and also storm-warnings, as issued by the British Meteorological Office for every district; each forecast being considered under the separate headings of "Wind" and "Weather," and the amount of success or failure is divided into four degrees—complete success, partial (more than half) success, partial failure, and total failure. In practice it is found that the percentage of any district varies but little from year to year, though, on the whole, there is a slow progressive improvement. The subjoined summary of weather-forecasts for the year ending March 31, 1882, may, therefore, be taken as a fair sample of the results usually attained by the Meteorological Office.

SUMMARY OF RESULTS.

District.	Percentages.				Total percentage of success.
	Complete Success.	Partial success.	Partial failure.	Total failure.	
Scotland, N.	39	42	14	5	81
Scotland, E.	35	43	15	7	78
England, N.E.	32	46	17	5	78
England, E. ...	33	44	17	6	77
Midland Counties ...	31	46	18	5	77
England, S.	35	46	14	5	81
Scotland, W.	30	44	19	7	74
England, N.W. ...	32	44	17	7	76
England, S.W.	34	42	18	6	76
Ireland, N.	36	44	14	6	80
Ireland, S.	35	41	16	8	76
Summary	34	44	16	6	78

By this it will be seen that the complete or partial successes amount to seventy-eight per cent., varying from seventy-four per cent. in the west of Scotland to eighty-one per cent. in the north of Scotland and south of England.

Checking Forecasts.

It might appear at first sight that when a forecast had been issued, it would be the simplest thing possible to check it, and to say whether it had been successful or not.

In practice, however, it is very different, as will be seen from the following remarks. The difficulty arises from two sources—the local variation of wind and rain in the same district, and the difficulty of assigning a mechanical measure to such elements as a gale of wind or a rainy day.

For instance, some of the British forecasting districts are about two hundred miles by one hundred, and contain two or three hundred square miles. Even within this limited area considerable differences of weather may be experienced. It may blow a gale at Dover, and only a fresh breeze in London, though those towns are only seventy miles apart. The difficulties are even greater when we come to treat the British Islands as a whole.

For instance, suppose we want to test the truth of the popular saying as to the frequency of gales at the equinox, how are we to define what is a gale? Is it enough to prove the saying if a gale has been experienced in only one of the eleven districts, or must we report a gale from three or four districts at least, before we can say that a storm swept over Great Britain about such or such a date? It follows from these general considerations that the total success which is credited to any district will always be much better than if the records at any one station had been compared with forecasts issued to the district in which it lay. Every office checks its own forecasts by its own method, so that the relative percentages of success which we shall give hereafter cannot be strictly compared. They are, however, very good approximations to the truth.

German.

We will pass from the consideration of winter and autumn gales, which move in an easterly direction, to the very different state of things which brings thunder and rain to Central Europe during the summer months. We have, therefore, selected an illustration of a partially successful set of forecasts issued by the Deutsche Seewarte at Hamburg on August 13 and 14, 1880. This will be a very typical example of rain with secondaries, and of the apparent independence of rain on the barometer. In Fig. 95 we give a synoptic chart over the greater portion

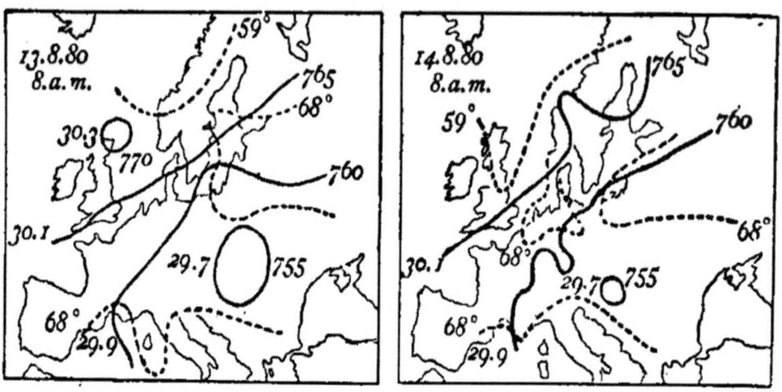

FIGS. 95 and 96.—Partially successful forecast (Germany).

of Central Europe at 8 a.m., Hamburg time; and in Fig. 96 a similar chart for the succeeding morning. The broad features of these two days is very simple. An anti-cyclone rests over Great Britain, while a shallow cyclone is moving westwards up the middle valley of the Danube. When we come to look at the movement of the isobars between the first and second day, we find that the position of the line of 29·9 ins. (750 mm.) has scarcely altered, but that the isobar has become looped up into secondaries.

The result of all this on the weather was to produce rain and thunder with very little wind, and insignificant changes in the reading of the barometer at any station.

The following forecasts were issued from the office of the Deutsche Seewarte at Hamburg for these two days:—

"Prospects for the weather of August 14, 1880, in Germany.—*General.* Continuance of the changeable weather, with precipitation, and light to fresh wind—in the north, mostly northerly; in the south, mostly westerly to northerly, with a temperature little changed or else falling. Here and there thunderstorms.

"Prospects for the weather of August 15, 1880, in Germany.—*General.* Rather warm—in the west for the most part bright weather; in the east overcast weather prevalent, with light wind. Inclination to thunder."

The cyclone which we see on our first chart lying over Hungary had been moving slowly for two or three previous days up the valley of the Danube river, and the above forecasts are evidently based on the supposition that the motion of the depression will continue in the same direction. The forecaster who is skilled in the meteorology of Germany knows both the kind of cyclone which moves westwards and also the kind that will develop rain and thunder, but he cannot tell exactly where the rain will be heaviest, nor whether the intensity will increase.

For this latter reason the above forecasts do not sufficiently indicate the very heavy rain and disastrous floods which occurred in Austria and South Germany during the period in question.

J. Hann (LXXXII. Bunde Accad. d. Wiss. II. Abbth. Nov., 1880) has made the weather of this period the subject of a special memoir. He finds the following dates for the heaviest rainfall:—August 11, Siebenburg and

South-East Hungary; 12th, all Hungary, Schlesien, Nieder Ostereich; 12th and 15th, Ober Ostereich, east of South Bavaria; 13th, west of South Bavaria, Bohemia, Saxon Erzgebirge; 14th, North Tyrol and Pinzgau; 15th, second maximum, Salzkammergut, West Schlesien, North Bohemia. That is to say, that on the whole the position of greatest rainfall travelled westwards with the primary cyclone. Some, but not all, of the rain was accompanied by thunderstorms. His investigations were mostly from the point of view which connects rain with the motion of the barometer, as observed at any one station. The results which he obtains are most striking illustrations of the principle we have so often alluded to, that the rain of secondaries is out of all proportion to the barometric changes as recroded by a solitary observer; and that the position of heaviest rainfall cannot be given from an inspection of the isobars, as in the case of primary cyclones.

The conclusion which he arrives at as regards these two points are as follows:—

"The appearance of a barometer *minimum* in Hungary occasioned enormous and extended precipitation on the west and north-west sides of this barometer depression. A reaction of this precipitation on the position of the centre of the depression is scarcely perceptible.

"Also the general distribution of pressure (the form of the isobars) shows no relation to the area of the intense precipitation.

"We find, therefore, through the investigation of the relative lowest barometer reading in its behaviour to rainfall, that our former conclusions are confirmed.

"A relation between barometer change and rainfall is scarcely obvious, and the conclusion is justified that the barometer fall, in the first instance, does not depend upon rainfall, and especially is not perceptibly influenced by the last."

Seewarte Success.

The following is the result of the tests of the general (*allgemein*) weather-forecasts published by the Deutsche Seewarte at Hamburg in 1882. The total percentage of success is credited with half the partial success.

Each element ...	Weather	78 per cent.
	Wind	75 "
	Temperature	78 "
General	Success	69 "
	Partial success ...	15 "
	Failure	15 "
	Total of success ...	77 "

United States Forecasts.

We will now give some illustration of forecasts and results in the United States and Canada. As an example of successful forecasting, let us look back at Figs. 30–35, in which we gave very detailed charts of the wind and weather in the United States on January 20 and 21, 1873.

Figs. 31 and 34 give the isobars, wind, and weather on the 21st, 4.35 p.m., Washington time. The cyclone which we found there over the Middle States had travelled in an east-north-east direction since morning, as we see by reference to the preceding chart (Fig. 30).

The subjoined forecast was evidently based on the

idea that the cyclone would continue to move in the same direction, so as to pass over the New England States, and that the Rocky Mountains cyclone would develop and advance eastwards. That is to say, that in the New England States the wind would shift to the north and west with a rising barometer, falling thermometer, and clear sky of the rear of a cyclone; that in the Middle States somewhat similar weather would be experienced: but that from Tennessee, northwards over Ohio, the wind would shift to the south and east, with a rising temperature and cloudy sky, from the action of the front of the new cyclone.

In the result, by 11 p.m. the same day (see Figs. 32 and 35), the first cyclone moved as expected, and the forecasts were a complete success in the New England and Middle States.

The new cyclone did not, however, advance as anticipated, and the southern anticyclone increased in size. Hence, from Tennessee, northward over Ohio, though some south and east wind was experienced, the weather remained fine, and the temperature fell from the radiation of the anticyclone, instead of rising for the cyclone. Hence the forecast was only partly verified. Nothing could show more clearly the difficulties of a forecaster than this example. He was unquestionably justified in expecting the advance of the new cyclone, but he was baffled by one of the endless shifts which accompany the growth of cyclones.

Turn now to the charts for the next day, which we gave in Figs. 42–44, to illustrate the nature of diurnal temperature-variations. There we find by 11 p.m. that day (see Fig. 42) the second cyclone had advanced very much as had been expected, only more slowly. The increase of the anticyclone over the Southern States on the first day was due to that gathering-up pressure which we have seen so often precedes the full development of an incipient cyclone, but could not have been forecast in our present state of knowledge.

The most unsuccessful portion of the forecast related to the weather in the north-west. The probabilities were given for "winds shifting to northerly and westerly, with rising barometer, falling temperature, and clearing but partly cloudy weather." This was based on the supposition that the new cyclone would be small and follow nearly the track of the preceding depression. The prediction was not justified by the result, but shows very clearly the scope of individual judgment. The following is an exact copy of the published synopsis and probabilities:—

"Washington D.C.
"Tuesday, January 21, 1873, 4·35 p.m.

"*Synopsis.*

"The barometer has continued falling, with rising temperature from Florida to the Middle and New England States, the lowest being central over the Lower Lake region, where fresh and brisk variable winds and rain and snow are now prevailing. Cloudy weather, rain, and fresh and brisk southerly to easterly winds are now prevailing from North Carolina to New York and New England, excepting light snow over northern part of latter. Generally clear weather from South Carolina to

"General Remarks as to Verifications.

"The above 'Probabilities' were generally verified, except 'from Tennessee, northward over Ohio and southern portions of Michigan and Wisconsin, winds shifting to southerly and easterly, cloudy weather, and possibly light rain,' and 'for the North-west, clearing but partly cloudy weather,' partly verified; 'from Tennessee, northward over Ohio and southern portions of Michigan and Wisconsin, rising temperature,' and 'for North-west, westerly winds,' not verified."

The foregoing example will fully illustrate the great advantage which the New England States possess in the ease with which cyclone-depressions can be traced before they reach the eastern seaboard. To this circumstance, and to the energetic management of the Signal Office, we may fairly attribute the high percentage of success which is achieved in the United States. The following table gives the percentages of success both of weather-forecasts generally, and of special storm-warnings:—

Year.	Weather forecasts. Per cent.	Storm warnings. Per cent.
1872	76·8	70
1873	77·6	—
1874	84·4	75
1875	87·4	76
1876	88·3	77·3
1877	86·2	78·9
1878	88·4	75·9
1879	90·7	79·9
1880	90·3	83·4
1881	88·7	83·3
1882	88·2	83·0

These results, like those of the British Meteorological

Office, differ little from year to year, but still show a slow progressive improvement. A few details of the methods employed in the United States Signal Office will be very useful to show the practical conditions of weather-forecasting. They are compiled from the publications of that office. From reading in the morning papers the "Synopsis and Indications" for the day, no one not initiated in the method of preparing them would suspect the magnitude of the work involved in their elaboration. The study requisite for the tri-daily press reports includes the drafting of seven graphic charts, exhibiting the data furnished by the simultaneous reports telegraphed from all the stations, about seventy-five in number. These charts are—

1. A synoptic chart of pressure, temperature, wind's direction, and velocity, the state of the weather, and the kind and amount of precipitation.

2. A chart of dew-points at all stations.

3. A chart of the various cloud-conditions prevailing at the time over the United States. The cloud-areas—each form of cloud represented by a different symbol—are outlined, and each one is distinguished. The appearance of the western sky at each station as observed at sunset, which affords a strong indication of the weather to be anticipated for the next twenty-four hours, is also marked in this chart.

4. A chart of normal barometric pressure and of variation of the actual from the normal pressures.

5. A chart of actual changes of pressure occurring, showing separately the fluctuations of the atmosphere during the previous eight and twenty-four hours.

6. A chart of normal temperature and the variations of the actual from the normal temperature.

7. A chart of actual changes of temperature in previous eight and twenty-four hours.

All these charts have to be made out, and the mass of data which they embody to be sifted and analyzed, preliminary to the preparation of every bulletin. Armed with this charted material, the officer preparing the indications proceeds to compile the "Synopsis and Indications," and issue the necessary storm-warnings. The average time which elapses between the simultaneous reading of the instruments at the separate stations, and the issue of the forecasts, is one hour and forty minutes.

Canadian Success.

The following particulars of the success obtained by the Canadian Meteorological Office are also interesting, for they give a percentage almost identical with that of the neighbouring States, though obtained by a different organization.

Storm-Warnings.

The percentage of warnings verified was—

1877	69·0
1878	78·3
1879	83·0
1880	82·8
1881	85·0

This table, like the others, shows progressive improvement. The only year for which we have the results of general weather-forecasts is 1881. Then the general percentage of complete success was 82·3 for the whole Dominion, while the proportion of partial and complete success rose to 90·2 per cent.

Australian Forecasts.

We will conclude this chapter with an example of forecasting in Australia, for which we are indebted to Mr. R. Ellery, of the Melbourne Observatory. This will be a valuable illustration of the universality of the general principles we have already laid down. But, first, let us say a few words on the general character of Australian weather. The weather of that great island continent has, like every other country, peculiarities of its own, subservient to the great principles common to all the world.

The same general distribution of pressure holds good there as elsewhere:—a low-pressure zone near the equator; a sub-tropical belt of anticyclones; an area of low pressure in the temperate zone, incessantly traversed by an endless series of cyclones. Within this latter area the same seven fundamental forms of isobars are perpetually reproduced; and the same kind of sky is developed in the equivalent part of each shape of isobars, and the same prognostics hold for good or bad weather, as in the northern hemisphere. Only the sequence of the wind as it veers during the passage of a cyclone is the opposite to that in the opposite hemisphere, because the rotation of the wind round the central vortex is in a contrary direction. For instance, we find the characteristic dirty sky and muggy heat of a cyclone on the right or equatorial front in the northern hemisphere, with wind veering from south-east to north-west, while in Australia we find the equivalent weather in the left (there also the equatorial) front of the depression, with wind beginning at north-east and going round to south-west.

After these explanations, we can readily understand the principles on which the following Australian forecasts were issued by the Government Observatory in Melbourne. Let us look back at Figs. 38 and 39, in which we give the isobars and winds over all Australia on November 20 and 21, 1884.

In the first chart (Fig. 38), we see the southern edge of the equatorial zone marked by the isobar of 29·9 ins. over Northern Australia; the edge of a great tropical anti-cyclone lies over Queensland; and the fragment of a temperate cyclone covers the great Australian Bight. The wind is light and variable at all the northern stations, but rotates round the cyclone in the usual manner. Now, from the peculiarities of Australian weather, the north-east or north winds in front of a cyclone of such moderate intensity are fine, though sultry, but occasionally a small thunderstorm develops, especially near the trough. The cyclone, as a whole, will certainly move towards the east, and the wind at every station will veer according to the universal rule.

Hence the following forecasts were issued at 3 p.m. "South" and "North" refer to those portions of the colony of Victoria only, and not to the whole of Australia:—

"South. Fine, sultry weather, with northerly tending to westerly and south-westerly winds, with thunder showers.

"North. Ditto. Ditto."

Now, if we look at Fig. 38, we see that the general anticipations have been fulfilled. The depression has moved towards the east, and the wind in Victoria gone round to west and south-west. But a new anticyclone has made its appearance over Western Australia, the cyclone has increased in depth, and thrown out a V-depression into the col between the two anticyclones. Hence the intensity has increased, and the weather is more unsettled on the second than had been expected on the first day.

INDEX.